THE LAST DEGREE

FROM TIPPING POINTS TO TRANSFORMATIVE
ACTION—UNITING DATA, POLICY, AND GRASSROOTS
POWER FOR A SUSTAINABLE FUTURE

MADHUSUDHAN ANAND

INDIA · SINGAPORE · MALAYSIA

ISBN
Paperback 979-8-89699-940-9
Hardcase 979-8-89929-286-6

The Last Degree

"Global warming isn't a prediction. It is happening."

– James Hansen

"I never imagined that the subject I would begin to study had such huge consequences."

– Syukuro Manabe

Contents

My Previous Works 7

A Note of Gratitude 9

Preface 11

A Snapshot of My Findings 21

Introduction 31

1. The Brink of Cascade 35

2. The Human Factor 73

3. Policy, Diplomacy, and Global Cooperation 103

4. Voices of Change 123

5. Turning the Tide 139

6. Mitigation Strategies and Solutions 153

7. The Role of Technology and Innovation 179

8. Envisioning the Future 195

9. Climate Jobs & Opportunities 219

10. Climate Data—The Backbone of Understanding 243

11. Climate Intelligence—Transforming
    Data into Action 259

12. Closing Note 273

References 291

My Previous Works

Resilience in Disruption: Thriving in the New Normal (First Book)

My journey as an author began with Resilience in Disruption: Thriving in the New Normal. Written at a time when the world faced unprecedented changes and uncertainties, it explores how we can adapt, overcome setbacks, and rediscover our strengths. The book underscores the importance of mental fortitude, emotional agility, and the capacity to innovate amid shifting landscapes—whether those shifts stem from market disruptions, global crises, or personal upheavals.

Boundless: Breaking Free from Limiting Beliefs (Second Book)

My second book, Boundless: Breaking Free from Limiting Beliefs, builds on the resilience framework by zooming in on personal transformation. This work delves into the internal barriers that often hold us back, revealing practical strategies to unlock hidden potential. From challenging self-doubt to cultivating empowering thought patterns, Boundless encourages readers to identify and dismantle any beliefs that keep them from reaching their fullest expression of who they are.

Connecting the Dots to This New Book

As you dive into my third book, you'll notice threads of continuity from my earlier works—especially the emphasis on adaptability, growth, and a forward-thinking mindset. Here, I expand that focus beyond the personal and professional realm to encompass the broader challenges we face collectively. Just as Resilience in Disruption encouraged us to pivot in unstable environments, and Boundless challenged us to transcend our internal limitations, this new book examines how we might harness resilience and possibility-thinking to address one of the most pressing global issues of our time: climate change.

If you are new to my writing, I invite you to explore these earlier books as they provide helpful context on navigating change, overcoming adversity, and freeing ourselves to innovate. Whether you're interested in personal development, leadership, or global challenges, each book offers a unique lens on how we can thrive—even in the face of uncertainty.

A Note of Gratitude

Writing this book has been one of the most difficult and rewarding experiences of my life, and it would not have been possible without the love, patience, and steadfast support of my family. To my wife, your conviction in my vision and consistent encouragement have helped me get through hard nights and early mornings. My sons, Mayur and Mayank, you are my inspiration and motivation. Watching your interest about the world just deepens my resolve to contribute to a healthier planet for you and future generations.

I want to express my heartfelt gratitude to the committed scientists and researchers behind NASA's PACE satellite mission, who introduced me to the tremendous possibilities of hyperspectral satellite imagery. Your pioneering work sparked my interest in how modern technology might provide new perspectives on the Earth's shifting climate. You demonstrated how, when used properly, data can help us understand both our planet's fragility and our own ability to protect it.

Finally, to the innumerable climate scientists and researchers throughout the world who work tirelessly—often behind the scenes—to unearth critical insights and solutions: your efforts power every page of this book. Thank you for pushing the boundaries of knowledge and working every day to protect the world we enjoy. Your passion reminds us all that optimism, when joined with action, may truly influence a better future.

Preface

Why did last summer set records for heat while this year has had continuous rains and threatening drought in other areas? Why does food cost twice as much from one season to the next, and how do pests or mosquitoes suddenly take over entire communities? Why can entire economies rise or fall based on the weather? Looking back, I realize I've always been intrigued by these questions, which revolve around the often-unseen powers of climate.

As a child, I noticed that certain summers were unbearable—thick, humid air pushing down on my chest—while others appeared abnormally chilly. I'd watch people around sneeze whenever the flowers opened in early spring and wonder why I complained of sudden headaches when a breeze blew in from a nearby industrial zone. Even then, I understood that something unseen was impacting our health, attitudes, and even the food on our plates.

Over time, I realized how these same dynamics influence everything from the pricing of produce at the local market to the spread of diseases like dengue or malaria, as well as the overall economics of entire countries. Farmers lose crops due to unanticipated heatwaves; bugs thrive in previously inhospitable locations; and supply chains choke or adapt to the weather's next turn. All of this led me to a realization: in addition to the business intelligence we rely on every day—such as anticipating stock prices or consumer behavior—a new, equally important type of intelligence is emerging. I refer to it as climate intelligence: real-time, hyperlocal data that reveals how the air we breathe, the temperatures we experience, and the humidity or pollutants we encounter directly influence our personal well-being and collective future.

This quest for climate knowledge did not inspire me to launch my startup, Ambee. But this whole climate thing started affecting me when my 6-month-old son and my family started facing its effects when what started as the common cold started affecting our lives to a matter of questioning our ability to survive and threatening to destroy everything we had created for ourselves. That led to me starting Ambee. What began as a basic effort to check local air quality—so that people in my neighborhood could breathe easier—quickly exposed a wider problem: dispersed, out-of-date environmental data and a lack of real-time insights into concerns such as pollution and pollen. I recognized that if I could figure out how to collect, analyze, and distribute this type of data, it would transform how we adapt to our ever-changing environment.

This book tells the story of how the fascination with the invisible forces around us grew into a quest to provide precise, actionable, and genuinely hyperlocal environmental data to communities all around the world. I ask you to join me in studying how climate intelligence may improve our health, economy, and the future of our planet—because once you see these invisible forces, you can't help but want to use them for good.

The journey began with an ambitious goal: to capture air quality insights at the neighborhood level. At the time, there was little equipment for monitoring pollutants, and those that did exist were either prohibitively expensive or only produced wide area averages. These averages didn't help a Mumbai woman decide whether to take her asthmatic child outside nor did they help a commuter in Los Angeles figure out how awful traffic-related pollution may be along the expressway on that particular day. The market sorely needed real-time, hyperlocal data on which regular people and businesses could make judgments. This is how Ambee began: with a team of people full of ideas eager to harness the potential of data science, remote sensing, and ground-level sensor networks. Our objective was to make environmental data more accessible and enable climate intelligence that had a real impact on people's lives.

As we began to refine our air quality algorithms, I became aware of a less evident but equally important factor: pollen. Many individuals view allergy season as a nuisance to be endured with tissues and medications. However, I observed a significant gap in real-time pollen measurement, particularly in India and other regions of the world. Meteorological organizations and private weather services frequently measure pollen counts in nations such as the United States, but pollen data is virtually nonexistent in India, where allergies and respiratory disorders are equally prevalent. Recognizing that air quality included more than just PM2.5, PM10, or gaseous pollutants, we chose to integrate pollen mapping into our solutions. Another challenge to tackle was how to generate comprehensive, street-level pollen information in cities without ground stations or reference data.

During these early days, we consistently heard the phrase, "That's impossible." Many people I spoke with, including specialists, mentors, and possible investors, were unconvinced that such granular data could be properly recorded on a large scale, let alone monetized or grown into a long-term business. Potential partners questioned if we could combine satellites, specialized sensors, meteorological models, and machine learning algorithms to produce a relevant report for someone's unique street. Even well-wishers advised us to set modest goals, such as measuring air quality in a few metropolitan regions or limiting pollen tracking to known allergen hotspots. However, the more we were instructed to limit our ambitions, the more determined we grew. There had to be a method to address this obvious gap in environmental intelligence. After all, climate change was not going to wait for technology to catch up; rather, the problem was escalating, bringing with it more frequent pollen surges, heatwaves, and pollution events that endanger public health.

Building from Scratch in India

Tackling the "impossible" required facing a practical reality: India lacked a pre-existing, high-resolution pollen record. Pollen monitoring programs have been in place in many developed countries for decades, with samples collected from authorized sites and analyzed under microscopes to identify specific plant species. However, outside of infrequent research programs, such networks were almost non-existent in India. We had to create a framework from scratch, learning to forecast pollen using plant biology, climatic data, and remote sensing images. This entailed determining which plants were flowering when, which wind patterns could transport pollen from one location to another, and how humidity levels affected pollen distribution. It was a crash education in ecology, meteorology, and data analysis. Our workplace tables were covered in scientific papers, satellite picture prints, and algorithms scrawled on whiteboards. Each minor discovery felt like a breakthrough, and each iteration of our model brought us closer to a viable answer.

Scaling Globally

Once we had proof of concept—street-level air quality and pollen data that worked consistently in a few test cities—another concern arose: Could this system be scaled internationally? The earth has numerous climate zones, diverse vegetation, and vastly differing pollution sources across continents. We wanted to investigate if our model could adapt to local settings, such as a tiny town in Germany or a large city in Brazil. We also didn't want to limit our coverage to India or a few countries. We intended to have a significant impact on global climate intelligence. We gradually increased our coverage, refining models for each new area, training them on local ground truth data, and partnering with partners around the world to get feedback. The day we saw our algorithms were estimating pollen and air quality in 160 countries—from urban centers in Southeast Asia to rural areas in North America—I knew we were on track to do the impossible.

Billions of API calls: An Endurance Game

Making these datasets available at scale was just half of the equation; the other half was dependability and speed. We needed to make sure that any third-party app developer, government agency, or business that used our application programming interface (API) could get accurate, real-time updates in milliseconds. This entailed creating an infrastructure capable of handling billions of API calls without fail. In the beginning, server breakdowns and latency issues were common problems. Each issue prompted us to reconsider our architecture: how to efficiently store, analyze, and send massive amounts of data in near real-time. We implemented cutting-edge cloud technologies, container orchestration, and microservices. Then, we refined and improved again, constantly expecting traffic to increase further. Over time, we developed a strong pipeline that could scale horizontally, matching the demands of the busiest allergy seasons and pollution spikes.

The complex interplay of data science and remote sensing was central to our endeavor. With NASA PACE (Plankton, Aerosol, Cloud, and Ocean Ecosystem) and other satellite missions, we suddenly had an abundance of atmospheric data. We could see aerosol distributions, cloud patterns, and even chlorophyll concentrations in seas, providing a more complete picture of Earth's changing environment. However, deciphering these raw satellite signals was no easy process. We needed to create machine learning models that could analyze these spectral fingerprints and compare them to ground station data, weather patterns, and vegetation indices. In effect, we were converting billions of pixels from orbit into useful information for a local community. This was a surprise for me: I realized that data communicates tales about planetary health and emerging climate concerns.

Conversations That Shaped Us

One of the most eye-opening aspects of this journey was speaking with folks who were directly affected by our findings. Mothers dealing with their children's allergies, city officials dealing with pollution ordinances, corporate sustainability leaders concerned about employee health, farmers concerned about changing rainfall patterns—these voices repeatedly reminded me that our work was more than just APIs, sensors, and lines of code. It was about making a tangible difference in someone's life, even if only slightly. For example, a large city's municipal board in Asia used our air quality alerts to develop a graded reaction action plan that limited construction activity and vehicle traffic on high-pollution days. During peak allergen seasons, a hospital network employed more specialists based on our pollen projections. These stories inspired our desire to improve and gave us the moral imperative to continue striving for greater accuracy.

We also learned how important the backing of investors and venture capitalists would be in getting us to the next level. Each round of funding brought not only cash resources but also a larger network of advisors and fellow entrepreneurs. However, the journey was far from straightforward or simple. Early on, we were repeatedly rejected, with many questioning the business sustainability of a hyperlocal environmental data startup—especially one that insisted on leaving a worldwide imprint or digging into specialist sectors like pollen data. It took several pitches, presentations, proof-of-concept demos, and real-world success stories to obtain the necessary buy-in. Looking back, I'm happy for both the "no" responses, which forced us to sharpen our vision, and the "yes" answers, which strengthened our commitment. Our collaboration with our investors has taken us further than we had anticipated.

As we dug further into air quality and pollen data, we became more aware of the full scope of climate intelligence. As the world grapples with mounting climate impacts—heatwaves, rising sea levels, and water scarcity—the lack of trustworthy and localized climate data stands out as one of the most significant barriers to developing effective responses. Many societies and economies rely on out-of-date baselines or incomplete weather models, leaving people vulnerable to sudden shocks. Seeing this gap, we expanded our data offerings to include climatic vulnerability indicators such as humidity profiles and UV indexes, as well as forest fire alerts and flood risk estimates. Our goal was to present as full a picture as possible, combining air quality, weather, pollen, and climate patterns so that decision-makers at all levels, from families to corporations, could act quickly and based on data.

This is why I wrote the book. After years of working with raw data, sensor arrays, satellite imaging, and on-the-ground obstacles, I realized that climate change is a tapestry of intertwined threads. What began as an attempt to fill a gap in real-time air quality data demonstrated how important each piece of the climate puzzle is. If we actually want to confront the world's most serious existential issue, we must understand how phenomena such as pollen blooms, heat extremes, air pollution, and oceanic changes are linked together. Air pollution cannot be properly addressed without taking into account shifting climate trends. Similarly, pollen surges are linked to changed plant lifecycles caused by rising temperatures and changing precipitation. By providing data that covers all interrelated elements, we positioned ourselves at the vanguard of climate intelligence—an ecosystem in which precise, real-time knowledge underpins all conceivable solutions.

My personal perspective evolved along with Ambee's capabilities. What began as a tech startup story—fast servers, powerful algorithms, and massive data—evolved into something far more philosophical and holistic. We are in the middle of a worldwide confrontation with our planet's boundaries, and data is one of the most powerful tools we have for guiding collective human behavior. However, data alone cannot solve the climate catastrophe; it must be transformed into intelligence. By intelligence, I mean informed strategies, community resilience plans, and policy frameworks that are actually responsive to local realities. Creating intelligence from raw data necessitates everything from user-friendly dashboards to community engagement, policy discussions, and academic collaborations. This bigger ecosystem is what I seek to highlight in this book.

A Snapshot of My Findings

Over the course of building what everyone said was impossible—global, street-by-street data on air quality, weather, and pollen—I have also been continuously learning about climate change. We hammered out solutions while grappling with shipping delays for sensor parts, negotiating data-sharing agreements with international agencies, and bridging knowledge gaps that required reading scores of scientific journals. The more we overcame these challenges, the more insights we gained about the fragility of our planet and the urgency of collective action. This book is my attempt to bring those findings into a cohesive narrative so that you—whether you're an environmental professional, a student, an entrepreneur, or just a curious reader—can see how it all fits together. You can read any chapter individually that interests you or you can choose to read sequentially. Every chapter is independent of the other. The chapters ahead lay out the structure of that story:

- **Introduction**: "Why Climate Change Matters." This sets the stage for the urgent reasons we need to pay attention to climate science and climate data. It also points toward the kind of solutions and insights you'll find in the rest of the book.

- **Chapter 1: The Brink of Cascade**. Here, I discuss climate tipping points—those critical thresholds beyond which we trigger large-scale, irreversible shifts. It's a concept that, at first, seems purely academic but becomes acutely real when you look at melting polar ice and the abrupt changes in ocean currents we've studied through data.

- **Chapter 2: The Human Factor.** Climate change is not just about weather patterns; it's deeply intertwined with human activities. I delve into how our industrial processes, agricultural expansions, and urban developments shape the climate. Pollen, air pollution, and greenhouse emissions all intersect in these human-driven endeavors.

- **Chapter 3: Policy, Diplomacy, and Global Cooperation.** After building data platforms, I learned firsthand how crucial global agreements—like the Paris Accord—are in shaping broader commitments. This chapter highlights the interplay between policy frameworks and real-world applications, informed by everything I've seen talking to environmental agencies, governments, and global investors.

- **Chapter 4: Voices of Change.** This chapter focuses on grassroots movements, innovators, and global partnerships that transcend political boundaries. It's about the human stories behind data—a reminder that climate solutions become unstoppable when local communities align with large-scale policies.

- **Chapter 5: Turning the Tide.** We look at scenarios for the future: from dire projections if we continue with business-as-usual to a more optimistic vision made possible through technology, activism, and enlightened policy. This is where you'll see how data can guide us in building more resilient cities and ecosystems.

- **Chapter 6: Mitigation Strategies and Solutions**. I narrow the lens to examine specific solutions—ranging from corporate responsibility initiatives to regenerative agriculture and ecosystem-based adaptation. The examples here stem from real projects that harness data to measure effectiveness and scale up their reach.

- **Chapter 7: The Role of Technology and Innovation**. As you might guess, this chapter is especially close to my heart. It looks at everything from VR-based empathy tools for climate awareness to cutting-edge renewable energy breakthroughs. Here, I share how our own technologies at Ambee fit into the larger narrative of global tech-driven solutions.

- **Chapter 8: Envisioning the Future**. This is where we explore cultural shifts, psychological barriers, and how we can nudge societies toward a sustainable future. It's not enough to have the data; we must change the ways we collectively think and behave.

- **Chapter 9: Climate Jobs & Opportunities**. This chapter explores how climate jobs, spanning renewable energy, green transportation, and sustainable construction, are driving systemic change. It highlights the need for workforce development to meet climate targets and ensure a just transition for communities.

- **Chapter 10: Climate Data—The Backbone of Understanding**. Here, I discuss about the role played by climate data in shaping our understanding of a changing world. From historical observations to advanced satellites and predictive models, it explores how data enables us to track, predict, and respond to climate challenges. The chapter also addresses gaps in accessibility and equity, emphasizing the need for collaboration to ensure data benefits everyone.

- **Chapter 11: Climate Intelligence—Transforming Data into Action**. This chapter examines the transformative power of climate intelligence in converting raw data into actionable insights. By leveraging tools like AI, machine learning, and predictive analytics, climate intelligence enables governments, businesses, and communities to make informed decisions that mitigate risks, improve sustainability, and adapt to a changing climate.

Why You Should Read On

While climate change is a gargantuan topic, often teetering between pessimistic doomsday warnings and overly simplistic "silver bullet" solutions, I hope this book provides a balanced perspective. You will see how local phenomena like pollen levels tie into huge global mechanisms like El Niño or shifting monsoons. You'll learn that no effort is too small—local policies can influence national agendas, and personal choices can shift market dynamics. You'll also discover that as large as the climate challenge is, the realm of climate data and intelligence holds immense promise. The knowledge that once seemed inaccessible or bewilderingly complex can now be harnessed by city planners, health professionals, farmers, and everyday citizens in ways we couldn't have dreamed of even a decade ago.

A Human Story, Not Just a Tech Story

I want to emphasize that this book isn't only for "techies" or climatologists; it's for anyone who breathes air, who notices changes in the environment, and who cares about how these changes affect our collective future. My journey with Ambee and climate data is fundamentally a human story—filled with excitement, doubts, triumphs, setbacks, and a relentless drive to make a difference. When I share stories about building high-resolution pollen forecasts where data didn't exist or collaborating with NASA satellites on advanced remote sensing, it's not just a story of technological wonder. It's a testament to our human capacity to innovate, collaborate, and push past what was once deemed unachievable for the sake of the planet and future generations.

Connecting the Dots

As you dive into the chapters, you'll see recurring threads: feedback loops in climate systems, the tension between economic growth and sustainability, and the interplay of local and global forces. I encourage you to follow these threads wherever they lead. The climate conversation is as interconnected as the systems it describes—fossil fuel dependency doesn't just increase CO2; it also amplifies heat waves, which then affect pollen distribution, which influences respiratory health. Similarly, a city adopting electric buses doesn't just reduce local pollution; it also sets an example for others, influences markets for electric vehicle components, and helps shift the overall policy conversation. Understanding these feedback loops is the first step toward taking meaningful action in your own sphere of influence.

Reflections on Hope and Fear

Climate change, more than any other global crisis, provokes a mix of dread and hope. We see record-breaking temperatures, devastating wildfires, unprecedented hurricanes—and it's easy to sink into despair. Yet, we also see extraordinary human innovations, youth-led climate strikes, entire nations pivoting to renewables, and local communities reclaiming degraded landscapes. The moral of my own journey is that while fear may alert us to danger, it is hope that propels us forward. If we rely solely on fear, we risk paralysis, but if we maintain hope anchored in solid data and pragmatic solutions, we can channel that energy into creativity and resilience. My greatest hope is that by the end of this book, you'll share in that conviction and perhaps become an advocate for climate intelligence in your own right.

A Collective Endeavor

Everything you'll read about here rests on a collective endeavor. Even the largest tech breakthroughs at Ambee arose from collaboration—developers fine-tuning algorithms, ecologists interpreting pollen samples, city governments piloting our solutions, and communities providing invaluable feedback. This synergy is a microcosm of what the entire world must do in the face of climate change. Governments, businesses, NGOs, and citizens must work together, each contributing their unique strengths. Our story is but one strand in a global tapestry of efforts aimed at slowing the pace of climate change, adapting to its inevitable consequences, and safeguarding the ecosystems that sustain life on Earth.

A Word on Imperfection

I won't pretend that everything is neatly resolved. Running a climate data company means encountering daily reminders of our planet's fragility and the gaps that still exist in scientific understanding. It also means grappling with business constraints, funding challenges, and scaling nightmares. But each year, we refine our methods, scale our models, and broaden our partnerships, bridging more of these gaps. We need continuous collaboration, ongoing research, and fresh perspectives to keep moving forward. That's another reason I put these thoughts into a book: to invite more people into the conversation and perhaps inspire fresh ideas, research collaborations, or policy shifts.

Looking Ahead

The future holds both peril and promise. On one hand, climate change is speeding up, with new extremes hitting communities faster than predicted. On the other, technological breakthroughs—from AI-driven climate modeling to next-generation renewable energy—are accelerating at breathtaking speed, offering real solutions if implemented thoughtfully. The good news is that we are not powerless. Data-driven insights can help farmers adapt their sowing cycles, allow doctors to prepare for allergy seasons, inform individuals when it's safest to exercise outdoors and guide policymakers in designing greener cities. In short, climate intelligence is a game-changer; we just need the will to integrate it into daily life.

Cultivating a New Mindset

A heartfelt wish of mine is that readers walk away with a renewed sense of agency. Climate change often feels like a problem too large for individuals to address, but that sense of powerlessness dissolves when we see how each cog in the machine can contribute. By building hyperlocal environmental datasets, for example, you empower local health clinics, municipal bodies, and individuals to make data-informed choices. By reading and sharing the knowledge in this book, you can spark conversations in your home, workplace, or city council, thereby shaping the narrative that fosters bigger structural changes. Awareness is contagious; once you see that your daily actions connect to global climate dynamics, it's hard to go back to complacency.

An Invitation

So, I invite you to read on. Let these pages expand your understanding of the climate crisis and the myriad ways we can address it, from individual lifestyle changes to sweeping policy reforms. Immerse yourself in the science, the technology, the diplomacy, and the human stories that fill these chapters. As you turn each page, consider how air quality and pollen data might relate to your own life. Maybe you have a family member who struggles with respiratory issues, or you've noticed the impact of rising temperatures on your local ecosystem. Perhaps you're an entrepreneur looking for ways to build a sustainable solution or a public servant aiming to craft stronger environmental policies. Whatever your background, know that your role in this global challenge is invaluable.

A Sincere Thanks

Before you embark on this journey through the book's chapters, I want to extend my gratitude. To my entire team at Ambee and all our partners worldwide: thank you for persevering through skepticism, hardware failures, and endless lines of code. To our investors: thank you for believing in a vision that many said was unattainable. To the scientists who paved the way in meteorology, ecology, and remote sensing, your work laid the intellectual foundation that we continue to build upon. And to every customer, city official, or concerned citizen who reached out to discuss environmental data: your questions and feedback shaped our approach, reminding us that our ultimate goal is to serve the common good.

Turning the Page

Now, with this preface drawing to a close, I hope you feel the gravity of our journey and the significance of the chapters to come. More importantly, I hope you sense that amid the seriousness of the challenge lies a reservoir of hope—a hope nourished by collaboration, scientific progress, and the triumph of human creativity over cynicism. As we delve into the science of tipping points, the complexities of global policy, the stories of local communities, and the promise of new technologies, let us keep alive the idea that it's never too late to act. Each of us holds a piece of the solution. Each of us can bring climate intelligence into our circles of influence, making the intangible tangible, the abstract actionable, and the impossible achievable.

In the following chapters, you will see how an initially small venture—founded on the simple wish to measure local air quality—grew into a data-driven crusade for climate intelligence that touches lives across continents. You will learn about our precarious position on the brink of the cascade, the human factor intensifying climate risks, and the global frameworks striving to rein those risks in. You will discover stories of pioneers, activists, and communities who are turning the tide, forging a path that leads not just to survival but to a potential renaissance of human ingenuity. Finally, you will leave this book equipped with knowledge of mitigation strategies, solutions, and technologies to help you navigate the evolving reality of climate change with confidence and clarity.

Thank you for choosing to spend your time reading this. May it strengthen your resolve, fuel your curiosity, and remind you that—even when everything seems daunting—we are not alone. We stand together, looking at the same sky, breathing the same air, and hoping for a better, more resilient future for all. Here's to charting that path forward, one data point at a time, and making certain that knowledge spurs us to meaningful, sustained action.

Introduction

Why Climate Change Matters

"The Earth does not belong to us; we belong to the Earth." Though spoken by Marlee Matlin in a different context, these words resonate deeply in our discussions on climate change. We find ourselves at a critical moment in time where a seemingly small yet powerful figure—just one degree—has the potential to reshape the way we live. But why does this minor increase in temperature hold such significance? How could something so subtle have such a profound impact on our daily lives and global stability? This book delves into the consequences of this seemingly small yet monumental shift.

As the Earth warms, we are witnessing glaciers melt, ecosystems being thrown out of balance, and changes to our living environments. Can we truly comprehend the science of climate change quickly enough to alter the course of our future? Are we prepared to harness our creativity and innovation to address the situation before it's too late? These questions affect not just scientists or policymakers but each of us who share this planet. Have you ever wondered how the changing climate might affect your own family, livelihood, or hometown?

This book is more than just a narrative of impending disaster; it offers hope and guidance on how we can reverse course. It's a call to action for everyone—scientists, activists, and ordinary citizens alike—to confront the climate crisis head-on before we reach the point of no return. Are we ready to answer that call? Are we prepared to take responsibility before it's too late?

Rachel Carson once said, "It is a peculiar circumstance that the sea, the cradle of life, is now imperiled by the actions of one of its inhabitants." These words, spoken more than half a century ago, still ring true today as we grapple with the looming threat of climate change. Carson's insights remind us of the delicate balance we have disturbed—a balance that may soon tip beyond our control. Will we continue on this path of destruction, or can we make the necessary changes to secure a future for the generations that follow?

Carson's insights remind us of the fragile balance that exists within Earth's systems, a balance we have gradually tilted towards a dangerous threshold. Our actions—often small, like daily energy consumption, food choices, or travel habits—are cumulatively pushing the planet towards this critical tipping point. *The Last Degree* symbolizes may be a milestone of synthesizing my own discovery on climate for me, but this is also a moment in time where human progress may either stumble into catastrophe or rise to meet the challenge with resilience and ingenuity. The choices we make today will determine the trajectory of our collective future. Will we continue down a path that leads to irreversible damage, or can we harness our capabilities to steer toward sustainability and renewal? The power lies in our hands, now more than ever. I will be exploring these pressing questions and more in this book as an active researcher in this domain.

This book aims to tackle the pressing issues head-on while offering practical solutions. It will help you understand the science of climate change, not just from a global policy perspective but through stories that make the crisis personal and relatable. As a father, I've faced the frightening reality of hidden pollutants when protecting my child, which pushed me to launch Ambee, a climate intelligence startup. These experiences highlight how climate change is not a distant issue—it's already affecting our health, our homes, and the very air we breathe. This book presents not just facts and data but the lived experiences that bring those numbers to life. It takes you through compelling data, advanced models, and real-world stories to emphasize the urgency of acting now.

But simply knowing the facts isn't enough. We need action. This book is also a testament to the creativity and innovation already happening around us. From solar panels on our rooftops to urban gardens sprouting in city centers, there are solutions within reach that can help turn the tide. We'll explore breakthroughs in green technology, rethinking how we design our cities and how individuals—people like you—are at the forefront of climate action. You'll meet the innovators transforming industries and shaping policies that guide us toward a greener, more sustainable future. The time to act is now, and each of us has a role to play—whether it's through reducing our carbon footprint, supporting climate-conscious policies, or simply becoming more aware of how our everyday actions can contribute to a larger movement. Together, we can make the difference.

Ultimately, *The Last Degree: Leading the Charge in the Climate Crucible* is a rallying cry for collective action. It delves into the psychological and societal barriers that prevent us from fully addressing climate change, exploring why we often hesitate to act, even when we know what's at stake. Through this exploration, the book presents a powerful case for overcoming these obstacles, and it invites you—yes, you—to become a part of the solution. This is not just a call for policy change or technological advancements; it's a call for personal commitment and shared responsibility. Together, we can build a partnership rooted in curiosity, hope, and a common goal of securing a sustainable future for all.

Imagine two futures. In one, the intricate web of life on Earth slowly unravels, leading to a world that is barren, inhospitable, and starkly different from the one we know today. In the other, humanity comes together, pooling our collective wisdom, empathy, and innovation to forge a new path—one that protects the planet and ensures a livable world for future generations. These are not distant, hypothetical scenarios; they are real choices that confront us right now. At this critical moment, we must ask ourselves: What kind of world do we want to leave behind? Will we be the generation that rose to the occasion or the one that turned a blind eye to the greatest challenge of our time?

I invite you to join me on this journey through science, challenges, and optimism. Together, let's take action now— before the final degree locks in our fate. The seas will endure, and the Earth will continue, but the balance of life and our place in it hangs in the balance. The future is ours to shape, and it begins with the choices we make today.

Let's lead the way in this climate battle because the stakes couldn't be higher, and the time to act is now.

The Brink of Cascade

Understanding Climatic Tipping Points

Imagine you're standing at the edge of a familiar riverbank, one you've known your entire life. You toss a small pebble into the water, and you watch as ripples fan out across the surface. The pebble seems insignificant, but its effects spread far beyond where it landed. This moment, as simple as it is, mirrors the way small changes in our climate can trigger far-reaching, sometimes catastrophic consequences. That's how I've come to think of climate tipping points—those seemingly small shifts that, once set in motion, ripple through ecosystems, weather patterns, and entire communities. Like that pebble, a small event can set off a cascade, altering the course of things in ways we might not anticipate until it's too late.

Now, imagine those ripples are actually dominoes—carefully arranged, each one representing an aspect of the environment, from glaciers and forests to ocean currents. A single nudge could trigger a chain reaction, setting off a series of irreversible changes. This is the essence of climatic tipping points—moments when Earth's systems cross a line, and everything begins to shift in ways that can't be undone. We're living in a time where these tipping points are more than just theories discussed in classrooms or research papers. They are alarms going off, warning us that we are dangerously close to losing control of the balance that sustains life as we know it.

Let me give you an example that always hits home for me as a parent. Imagine you're watching your child run towards the road. There's a moment—a split second—where you know you can still stop them, but after that moment passes, your ability to intervene diminishes rapidly. That's exactly what happens with climate tipping points. For instance, when Arctic ice melts, it exposes the darker ocean surface underneath. The darker surface absorbs more heat, which leads to even faster melting of the ice. This creates a feedback loop—a cycle that, once started, accelerates itself. Like that child running towards danger, if we wait too long, there's only so much we can do to stop the damage.

We don't have to look far to see this in real-world events. Just recently, we've witnessed the alarming melt of Greenland's ice sheet, which shed over 532 billion tons of ice in a single year—enough to flood the entire state of California with water. This rapid ice loss is not just about the Arctic; it is contributing to rising sea levels that threaten coastal cities across the world. Think about the severe floods in Pakistan in 2022 or the devastating wildfires that swept through parts of Europe and the U.S. in the same year—events that have been linked to climate shifts that are reaching dangerous tipping points. These are not isolated incidents. They are warnings of what's to come if we don't act swiftly.

But it's not just about reacting to these changes—it's about building resilience and understanding that the systems we depend on need to be prepared for disruption. Climate resilience is about more than just enduring a few hotter summers or rising sea levels; it's about designing our communities, economies, and daily lives to withstand and adapt to these sudden shifts. Think of it like constructing a strong levee around a coastal village, knowing the waters will rise, but being ready to protect what matters most. We may not always be able to stop the dominoes from falling, but with foresight and action, we can build the strength to absorb the shock and reimagine how we live in a world that's changing faster than we ever thought possible.

Take the Netherlands, for example. Faced with the constant threat of rising sea levels, the country has turned climate resilience into an art form. Innovative flood protection systems like storm surge barriers and floating homes have transformed their vulnerability into a model for how we can adapt to the new climate realities. It shows that while we may not be able to prevent every tipping point, we can certainly prepare ourselves to better withstand their impacts.

In short, tipping points are not just abstract ideas—they are very real, immediate threats. And resilience isn't just about hope; it's about action, about making sure that as the world changes, we're ready to meet it head-on with the tools, mindset, and urgency that this moment demands. This is where our personal stories intersect with the larger narrative of climate change. Each of us is part of that chain reaction, either pushing it forward or working to slow it down. The choice is ours to make, and the stakes couldn't be higher.

But it's not just about reacting to these changes—it's about building resilience and understanding that the systems we depend on need to be prepared for disruption. Climate resilience is about more than just enduring a few hotter summers or rising sea levels; it's about designing our communities, economies, and daily lives to withstand and adapt to these sudden shifts. Think of it like constructing a strong levee around a coastal village, knowing the waters will rise, but being ready to protect what matters most. We may not always be able to stop the dominoes from falling, but with foresight and action, we can build the strength to absorb the shock and reimagine how we live in a world that's changing faster than we ever thought possible.

In short, tipping points are not just abstract ideas—they are very real, immediate threats. And resilience isn't just about hope; it's about action, about making sure that as the world changes, we're ready to meet it head-on with the tools, mindset, and urgency that this moment demands. This is where our personal stories intersect with the larger narrative of climate change. Each of us is part of that chain reaction, either pushing it forward or working to slow it down. The choice is ours to make, and the stakes couldn't be higher.

Having worked with climate data for the last decade, accessing nearly every Earth observatory satellite, on ground weather, air quality and other ground stations and studying essential climate variables, I have witnessed firsthand the profound transformation of our planet. The data reveals a stark reality: the escalating intensity and frequency of wildfires are not just isolated events; they are part of a broader pattern that fuels climate change. In 2020 alone, global wildfires emitted 1.76 billion tons of carbon dioxide—an amount nearly equivalent to the entire annual emissions of the European Union. These fires create a chain where the more carbon dioxide released, the warmer the atmosphere becomes, leading to even more wildfires. It's a vicious cycle that is accelerating climate instability.

Consider the thawing permafrost in regions like Siberia and Alaska. Permafrost is important because it traps and stores carbon. A lot of carbon; an estimated 1,700 billion tons in fact. This is significantly more carbon than is found in the atmosphere.

An Image of permafrost in Siberia

It is helpful to understand a little about soil and its composition to better grasp this carbon-storing process. As these frozen layers of earth melt, they release massive quantities of methane—a greenhouse gas 25 times more potent than carbon dioxide. Recent studies estimate that permafrost regions could release up to 240 billion tons of carbon by 2100 if current warming trends continue. This methane release sets off another dangerous feedback loop, contributing to the rapid warming of the Arctic and beyond, pushing the planet closer to tipping points we may not be able to reverse.

Ecosystems like the Great Barrier Reef provide another stark example. Over the past five years, the reef has experienced three mass bleaching events, with over 60% of its coral affected. This isn't just a matter of lost biodiversity—it's a breach of our planet's boundaries. Coral reefs are vital to marine life and coastal communities, supporting over 25% of marine species and contributing $375 billion annually to the global economy through fisheries, tourism, and coastal protection. The loss of coral reefs will have cascading impacts, not just on ecosystems but on human societies as well.

Great Barrier Reef bleaching

We need to know what is happening for the sake of future generations, for the sake of our future and our children. Understanding these phenomena is not just an academic exercise—it's essential for cross-disciplinary, cross-border, and cross-generational discussions. These terms and processes are the distress signals of our planet, and by mastering the language of climate science, we equip ourselves to respond strategically. As we dive into climatic tipping points, we uncover an intricate web of interconnected processes that are unfolding with alarming momentum—momentum that may soon exceed our ability to adapt.

This accumulated knowledge isn't just theoretical. It serves as our guiding light, providing the foresight we need to anticipate, mitigate, and potentially halt the worst effects of these tipping points. At this pivotal moment, we face a crucial choice: will we allow these interconnected feedback loops to spiral out of control, or will we be the force that reshuffles the pieces toward a sustainable and balanced outcome? Join me as we navigate this critical juncture, armed with the data, insights, and collective will to shape a future where we rise to the challenge rather than succumb to it.

Historical Climate Shifts as Vital Warnings

As the dawn illuminates the fertile plains that were once bustling with early civilizations, we are prompted to reflect on the lasting influence of climate on human fate. The territories of ancient Mesopotamia, nourished by abundant rivers in the past, fell prey to aridity as nature's intricate equilibrium faltered. This narrative of ancient environmental transformations not only imparts historical insights but also issues a poignant caution for the present day.

Let's travel back to the end of the last Ice Age, approximately 11,000 years ago. During this time, the Earth was transitioning from the frozen period of the Pleistocene epoch, melting its icy covering as sea levels rose significantly. As glaciers retreated, land bridges became submerged, and coastlines changed rapidly in geological terms. Human civilizations, closely connected to these climate changes, had to adapt quickly or face extinction. This significant change emphasizes the rapid and relentless nature of historical climate shifts, reflecting the urgent challenges we face in modern times.

Advancing to the 1300s, Europe was gripped by the Little Ice Age. Crop failures, famine, and social unrest ensued as the climate grew colder and more unpredictable. The once prosperous Norse colonies in Greenland were deserted due to the harsh conditions, rendering survival impossible. This era was not merely coincidental; it served as a poignant illustration of the climate's influence on human fate, underscoring the significant repercussions of extreme climatic conditions on societal equilibrium and human existence.

Exploring more recent occurrences, the Dust Bowl of the 1930s stands as a somber reminder of the intricate relationship between human actions and climatic factors. Extensive farming practices and drought triggered massive dust storms that ravaged the Great Plains of the United States, uprooting thousands and compounding the challenges of the Great Depression. This incident highlights how human mismanagement can magnify natural climatic fluctuations and their severe repercussions.

Likewise, the Younger Dryas event approximately 12,000 years ago stands as a pivotal illustration of abrupt climate change. This era denoted an abrupt reversion to glacial conditions in the Northern Hemisphere, instigated by disturbances in ocean currents. It showcases the rapidity with which climate systems can transition, underscoring the likelihood of sudden changes within our present climate context.

From ancient shifts in Mesopotamia to more recent events, history not only imparts wisdom but also serves as a cautionary tale. The current climate challenges we encounter today echo the trials of the past, underscoring the enduring impact of our actions—or lack thereof. As we grapple with the consequences of a warming planet, we find ourselves at a critical juncture. One path leads to familiar territory marred by historical environmental negligence; the alternative, though less assured, demands innovation and determination to script a new chapter in our rapport with Earth.

The narrative that emerges underscores the imperative for societies to take the lead amidst climatic challenges. Our task is to draw upon historical insights to avert impending disasters. Through a comprehensive grasp of the catalysts and consequences of past climate variations, we empower ourselves with the wisdom to shape a sustainable tomorrow. In every shift from abundance to scarcity, from stability to chaos, the Earth offers us invaluable lessons in survival. The ancient city of Petra, a remarkable feat of engineering in the Jordanian desert, showcases human ingenuity in water conservation. However, even these sophisticated systems eventually yielded to climate changes, emphasizing the vital requirement for adaptive strategies in our current era.

The narrative of human history is intricately linked with climate, with each chapter guiding our future endeavors. Our responsibility lies in crafting a conclusion that signifies rejuvenation rather than decline and a legacy of responsible stewardship. The narrative we craft today embodies urgency and hope, envisioning a future where renewable energy supplants fossil fuels, deforestation transforms into reforestation, and sustainable practices reduce our ecological impact. Our present decisions shape whether past climate shifts act as precursors to decline or drivers of progress.

As we craft the upcoming chapter of 'The Last Degree: Leading the Charge in the Climate Crucible,' let us heed history's lessons as guides for the future. They prompt us to redirect our path, to shape a time when harmony with nature is reinstated. The climate crucible awaits our reaction. Will we meet the challenge, or will past events recur? The decision lies with us, and the moment to take action is now.

The 'Last Degree'—A Critical Threshold

In the overarching narrative of Earth's climate, every chapter relies on a fragile equilibrium. Currently, this equilibrium is precariously close to tipping over with just one more degree of temperature rise. What significance does this 'Last Degree' bear, and how crucial is it for the future of our planet?

The 'Last Degree' signifies a critical threshold in global temperature rise that, if surpassed, may result in severe repercussions for our climate. It denotes the tipping point where Earth's systems could veer beyond our influence, bringing about irreversible transformations. This marginal increase in heat, juxtaposed with Earth's extensive climatic timeline, holds the capacity to reshape our current understanding of life. The global average temperature, a composite of diverse regional and seasonal temperatures, provides a comprehensive assessment of the Earth's thermal condition. When scientists mention the 'Last Degree,' they typically allude to surpassing a 2°C elevation from pre-industrial times. This standard, established by the global community, signifies the maximum limit for controllable climate variations.

This crucial degree represents more than a mere figure; it delineates a boundary between two realms—one characterized by gradual, controllable alterations and the other marked by sudden, tumultuous changes. Within this threshold lie various factors: melting polar ice, escalating sea levels, severe weather phenomena, biodiversity decline, and transformations in agricultural output. Historically, Earth's climate has exhibited fluctuations; however, the notable acceleration in recent temperature increases, driven by human activities post-Industrial Revolution, has brought us dangerously near the critical 'Last Degree' threshold. Industrial practices, notably the extensive combustion of fossil fuels, have markedly raised atmospheric carbon dioxide levels, leading to intensified heat retention.

Exploring the scientific evidence reveals significant differences between a 1.5°C and 2°C increase. The Intergovernmental Panel on Climate Change IPCC offers precise projections:

- **At 1.5°C**: Sea levels are anticipated to rise notably, yet by a margin of 10 cm less compared to a 2°C scenario. This adjustment has the potential to avert the inundation of extensive coastal regions and alleviate the displacement of numerous individuals. The occurrence of severe weather events such as heatwaves and intense rainfall would be mitigated, lessening the repercussions on human health and agricultural activities. While coral reefs are expected to experience a significant decline, their total eradication is not projected.

- **At 2°C**: The implications are more severe. The accelerated melting of ice sheets in Greenland and Antarctica would elevate sea levels, heightening the risks of coastal flooding. There would be a substantial loss of biodiversity, potentially leading to the extinction of numerous species due to habitat depletion and changes in climatic conditions. Human health hazards, such as heat-related illnesses and the proliferation of diseases like malaria and dengue fever, would experience a notable rise.

The 'Last Degree' emphasizes the critical need to minimize global warming impacts. Meeting the 1.5°C target demands prompt and unprecedented worldwide actions to cut greenhouse gas emissions. This includes shifting to renewable energy, improving energy efficiency, and progressing in carbon capture and storage technologies. The distinctions between 1.5°C and 2°C of warming, although appearing minor, bear significant implications for Earth and its inhabitants.

- **Policy and Action**: Achieving the 1.5°C target necessitates prompt and unparalleled actions to diminish greenhouse gas emissions. This encompasses shifting towards renewable energy sources, improving energy efficiency, and deploying extensive carbon capture and storage technologies.

- **Socio-Economic Impacts**: Vulnerable populations, especially in developing nations, face heightened risks from exceeding these thresholds. Upholding climate justice and equity in climate action is imperative to safeguard those disproportionately impacted by climate change.

- **Long-Term Sustainability**: Adhering to the 1.5°C limit is crucial to upholding the resilience of natural and human systems. This limit offers a more feasible outlook for adaptation and diminishes the risk of activating catastrophic tipping points within the climate system.

Despite the unequivocal scientific evidence, persistent misconceptions hinder the necessary urgency in addressing climate change effectively. A common fallacy lies in underestimating the distinction between a 1.5°C and 2°C warming scenario. This oversight overlooks the fact that even a seemingly marginal rise in global temperatures can exponentially amplify the adverse impacts on our planet's climate system. The variance between these thresholds can entail significantly heightened risks of severe weather occurrences, increased loss of biodiversity, and more pronounced implications on human health and global economies.

Moreover, while some may minimize the need for immediate action by alluding to natural climate fluctuations in Earth's history, such comparisons overlook the unprecedented speed and scale of the changes caused by current human activities. Never in recorded history has humanity exerted such a profound impact on the global climate with such rapid consequences as is evident today. This highlights the exceedingly crucial state of our present circumstances, necessitating swift actions to mitigate the impacts of climate change.

As we navigate the narrative of the 'Last Degree,' we must ask ourselves: Will we heed the warnings etched into the fabric of our planet's history? Are we prepared to act collectively, drawing wisdom from both our triumphs and missteps? Imagine a world where the 'Last Degree' signifies not doom but a catalyst for transformation. Envision reforested lands, cities flourishing on clean energy, and oceans teeming with life—all achievable through our collective dedication to shift our environmental course.

In this critical juncture, grasping the importance of the 'Last Degree' is paramount, as is taking resolute action. The decisions made today will determine the legacy we bequeath to forthcoming generations. Will we overcome the hurdles presented by the 'Last Degree,' or will history recall us as the generation that failed to seize the chance to protect our planet? The stakes are significant, and the call to action is pressing. Let us rise to confront this challenge and guarantee that our chapter in Earth's enduring narrative is defined by valor, creativity, and an unyielding dedication to a sustainable world.

Domino Effects in Climate Systems

Within our planet's intricate climate system, every element is interconnected, creating a sophisticated matrix of environmental balance crucial for sustaining life as we understand it. At the brink of the Last Degree, our attention turns to the ripple effects - a chain of tipping points that, when set in motion, may spark an irreversible sequence of transformations, pushing us towards an uncontrollable climate scenario. Remember the domino effect we discussed earlier? Each piece symbolizes a vital element of Earth's climate, from the melting Arctic ice sheets to the coral reef bleaching and from permafrost thawing to Amazon rainforest depletion. The collapse of one can expedite the downfall of the next, occurring rapidly in succession, resulting in profound impacts on us.

The interconnectedness of tipping points is crucial. Scientists advise that surpassing one threshold could significantly raise the chances of surpassing others. With the retreat of Arctic ice, the dark ocean waters that remain absorb more heat, hastening global warming—an occurrence recognized as the albedo effect. Simultaneously, as permafrost thaws, vast amounts of methane, a potent greenhouse gas, are released, further worsening warming patterns. If the Amazon were to transition from a carbon sink to a carbon source, its trees would emit carbon rather than absorb it, intensifying global warming even more.

Here's a breakdown of nine key tipping points that could be triggered by climate change and how they could impact the planet.

1. Atlantic Meridional Overturning Circulation AMOC Collapse

This circulation in the Atlantic Ocean is crucial for regulating climate patterns, especially in Europe and North America. If global warming continues to weaken it, we could see major shifts in weather patterns, including colder winters in Europe and disruptions to the monsoons. While a full collapse might be irreversible, reducing global carbon emissions could stabilize it and delay crossing this threshold.

2. West Antarctic Ice Sheet Disintegration

The West Antarctic ice sheet contains enough ice to raise global sea levels by several meters. Rising ocean temperatures could lead to disintegration, and once this process starts, it's nearly impossible to stop. While we can't reverse the melting, slowing emissions can prevent a catastrophic collapse.

3. Greenland Ice Sheet Disintegration

Similar to the West Antarctic ice sheet, Greenland's ice is melting rapidly. If it disintegrates, it could contribute up to 7 meters of sea-level rise. Once this process accelerates, it is difficult to reverse, but reducing emissions now could help slow the process.

4. Amazon Rainforest Dieback

The Amazon, often referred to as the "lungs of the Earth," is at risk of dying back due to deforestation and climate change. If enough forest is lost, it could transition into a savannah-like state, releasing vast amounts of stored carbon into the atmosphere. Reforestation efforts and reducing deforestation could help restore parts of the ecosystem, but once a certain threshold is crossed, much of the damage could be permanent.

5. Boreal Forest Shift

The boreal forests of North America and Eurasia are increasingly threatened by wildfires, pests, and warming temperatures. If they shift northwards or convert into grasslands, they would release significant carbon. While some of this change may be irreversible, improving forest management and reducing emissions could slow the damage.

6. Permafrost Thawing

Permafrost in the Arctic is thawing, releasing methane, a potent greenhouse gas. This process creates a feedback loop that accelerates warming. While thawed permafrost cannot refreeze easily, cutting emissions and adopting carbon capture technologies could limit the release of additional greenhouse gases.

7. Coral Reef Die-Off

Rising ocean temperatures and acidification are causing mass coral bleaching events. Once coral reefs die, they take decades or centuries to recover, if at all. However, reducing carbon emissions and protecting marine environments could help slow the rate of bleaching and give corals a chance to adapt.

8. West African Monsoon Shift

The West African monsoon system is crucial for agriculture and water supply in the region. A shift in its patterns due to warming could cause severe droughts and flooding. While difficult to reverse, adaptation strategies like improved water management and sustainable farming could mitigate the impacts.

9. Indian Monsoon Shift

Similar to the West African monsoon, the Indian monsoon is at risk of shifting due to climate change. This could devastate agriculture and displace millions. Efforts to reduce greenhouse gas emissions and implement sustainable agricultural practices can reduce the severity of these changes.

The ramifications are significant. Escalating sea levels engulf coastal urban areas, leading to the displacement of millions due to climate changes. Intense weather occurrences become more regular and intense, causing disruptions to food supply chains and economic structures. Biodiversity declines as species face challenges in adjusting to swiftly changing circumstances, consequently compromising the ecosystems vital for human existence. The melting of Arctic sea ice serves as a crucial indicator of climate change, exerting extensive influence on global weather patterns and ecosystems. This phenomenon illustrates how disturbances in one facet of the climate system can instigate alterations throughout others, significantly affecting our global environment.

Arctic sea ice functions as the Earth's reflective shield, redirecting sunlight into space to regulate the polar regions and global temperatures. The albedo effect, pivotal for temperature equilibrium, hinges on this process. Melting sea ice unveils darker ocean water that absorbs more sunlight, hastening the warming trend. This feedback mechanism not only amplifies ice depletion but also heightens the impact of global warming.

The jet stream, a swiftly flowing current of air in the upper atmosphere, is influenced by the temperature variance between the frigid Arctic and the warmer mid-latitudes. With Arctic amplification accelerating warming in the Arctic region, the temperature gradient decreases, resulting in a more meandering and unpredictable jet stream. This phenomenon can trigger prolonged weather patterns, such as extended periods of extreme heat, cold, or precipitation, carrying significant consequences like the severe winter storm in Texas in February 2021 and the intense heat waves experienced in the Pacific Northwest during the summer of 2021.

As Arctic ice diminishes, it leads to the dilution of ocean water, potentially causing disruptions to key currents such as the Atlantic Meridional Overturning Circulation AMOC. A deceleration of the AMOC may result in cooling in parts of Europe and North America, as well as modifications to global rainfall patterns, impacting agriculture and water resources. The shifts in ocean salinity and temperature further influence marine ecosystems, leading to habitat disturbances and alterations in food sources.

To address the risk of cascading climate impacts, a comprehensive strategy encompassing mitigation, adaptation, and innovation is imperative.

- **Mitigation**: We must significantly decrease greenhouse gas emissions. This entails shifting towards renewable energy sources such as solar and wind, improving energy efficiency across various sectors, and safeguarding and rehabilitating forests and other natural carbon sinks.

- **Adaptation**: Developing resilience in communities highly susceptible to climate change is paramount. Enhancing infrastructure to endure severe weather, cultivating drought-resistant crops, and establishing early warning systems for natural disasters are all pivotal measures.

- **Innovation**: We should allocate resources to research and develop new technologies capable of capturing and storing carbon from the atmosphere or reflecting sunlight to cool the Earth.

Implementing these solutions necessitates international collaboration and coordinated efforts.

- **Policy Initiatives**: Policies should be implemented to encourage the transition to a low-carbon economy, including measures like carbon pricing and subsidies for renewable energy sources.

- **Financial Support**: Financial mechanisms need to be established to bolster climate adaptation initiatives in developing nations, guaranteeing inclusivity across all communities.

- **Public Engagement**: Education and communication campaigns are essential to enhance public awareness and engagement, empowering individuals to participate in climate action initiatives.

Evidence of the efficacy of such interventions is evident in historical achievements, such as the swift proliferation of renewable energy in specific nations, substantial reforestation initiatives, and the increasing urban centers pledging towards carbon neutrality. These stand as guiding lights, showcasing the feasibility of change. Nevertheless, alternative approaches warrant consideration. Advocates propose geoengineering - purposeful interventions in the Earth's climate system to mitigate global warming. Methods such as solar radiation management or ocean fertilization are controversial, carrying notable risks and uncertainties. Given the potential tipping points, should these strategies be viable options?

As we contemplate these potential solutions, let us not overlook the impact of individual actions. Whether it involves waste reduction or backing sustainable practices, each decision we make plays a part in a broader collective endeavor to prevent the cascading effects of climate change. What course of action will we take to address the climate crucible? Are we prepared to demonstrate the necessary courage and commitment to implementing solutions that can avert the environmental collapse chain reaction? Or will we passively observe as the consequences unfold, each impact reverberating until the resulting discord overshadows the balance of life on Earth?

The solution resides within each of us. The moment to take action is upon us, before the initial domino's descent, initiating an irreversible chain of events. Let us take the lead, steering away from the peril of an unchecked climate, and towards a future where stability reigns, preserving the delicate equilibrium of our world.

Polar Ice and Permafrost Peril

As the light of dawn illuminates the Arctic expanses, it unveils a landscape experiencing profound transformation. The previous silence, punctuated by the occasional crack of ancient ice, now echoes with the continuous drip of melting glaciers—a solemn indication of notable climatic changes.

In remote environments, polar ice caps and permafrost play pivotal roles. The luminous white glow of the ice reflects solar energy back into space, while the permafrost, though less apparent, securely stores vast amounts of organic carbon. Collectively, they function as stewards of Earth's temperature, upholding a fragile equilibrium that supports life. However, with the planet's rising temperatures, these icy guardians are receding. Polar ice caps are diminishing at concerning rates, and the thawing permafrost is releasing methane and carbon dioxide, potent greenhouse gases that further intensify atmospheric warming.

Addressing this challenge is as multifaceted as the issue itself. Strategies range from international policy agreements targeting the reduction of carbon emissions to localized initiatives for monitoring and adapting to these shifts. Technological advancements enable scientists to gain deeper insights into the intricacies of ice melt and permafrost degradation using satellite imagery, climate modeling, and field studies. These endeavors, though ongoing, offer a ray of hope amid grim forecasts, enhancing our climate models and informing impactful interventions.

Every fraction of a degree in temperature rise that we prevent has the potential to decelerate the rate of ice loss and permafrost thaw. This realization highlights the significant implications, not only for the Arctic and Antarctic but for the entire globe. The interconnected nature of our planet dictates that melting ice caps contribute to sea level rise, impacting communities situated thousands of miles away. Moreover, the thawing permafrost not only releases greenhouse gases but also destabilizes land, thereby affecting infrastructure and ecosystems in northern latitudes.

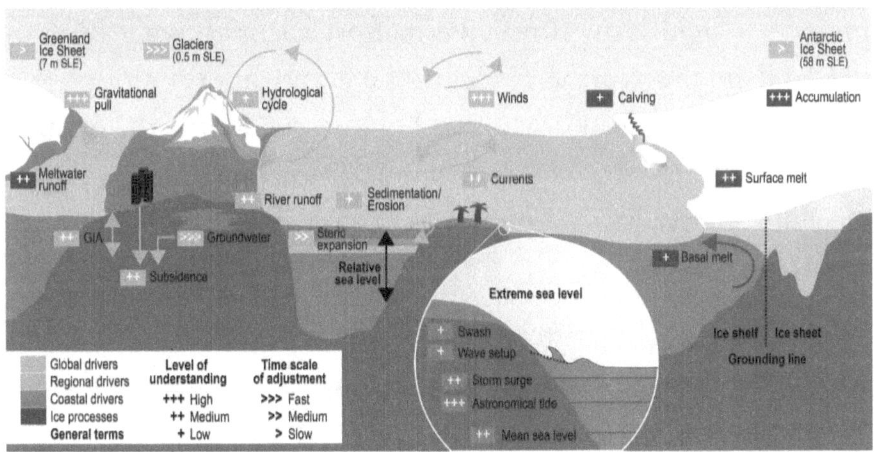

Visual aids, like graphs demonstrating the correlation between increasing temperatures and ice loss or maps outlining the possible expansion of permafrost thaw, are instrumental in conveying the pressing nature of the matter. These visuals function not only as educational resources but as poignant reflections of the challenges at hand. The melting of polar ice and permafrost due to global warming carries significant implications for global sea levels and methane emissions. The Greenland ice sheet alone experienced a loss of approximately 3.8 trillion tons of ice from 1992 to 2018, resulting in a sea level increase of around 10.6 millimeters. A complete melting of the Greenland ice sheet could elevate global sea levels by about 7 meters. Similar outcomes are anticipated from the melting of the West Antarctic ice sheet, potentially causing an additional sea level rise of approximately 10.5 meters.

In addition to ice melt, the heating of ocean waters contributes to sea level rise through thermal expansion. This phenomenon results in water expansion, consequently increasing sea levels. According to current projections, a sea level rise of 0.28 to 1.01 meters is anticipated by 2100, varying based on future greenhouse gas emission scenarios. The thawing of permafrost presents a significant threat. Permafrost spans roughly 24% of the land in the Northern Hemisphere and houses around 1.5 trillion metric tons of organic carbon, nearly double the existing atmospheric content. Upon thawing, it emits methane, a potent greenhouse gas, as a result of microbial decomposition in anaerobic conditions.

Evidence of these alterations is already apparent in locations such as the Lena River Delta in Siberia, where methane emissions have notably risen. These emissions have the potential to initiate a feedback loop that hastens global warming, resulting in additional thawing and increased greenhouse gas emissions. This cycle poses a threat to the disruption of climates and ecosystems on a global scale.

As I consider the upcoming phases of our journey, I reflect on the narrative that sparked the creation of Ambee. It was the pursuit of actionable information crucial for my son's well-being, which now motivates our endeavors in addressing the climate crisis. Just as knowledge empowered me to ensure my child's safety, it enables us all to protect our planet. Have we not gleaned from historical narratives and scientific lessons that inaction aligns with disaster? Can we mobilize the collective will to enact essential measures, or will we hesitate, thereby compelling future generations to navigate a world molded by our hesitance?

The decisions we make today - whether to innovate, adapt, and mitigate - hold significant importance. They will shape the legacy we leave for future generations. Will we take decisive action to uphold the delicate balance of our planet, or will we observe the consequential effects of our inaction reshaping the world? As we consider the significant transformations in the Arctic and Antarctic regions, we are prompted to reflect on the interconnectedness of our global climate system. May our legacy reflect our proactive approach to overcoming challenges, evolving into architects of a resilient, flourishing world rather than mere observers of catastrophe.

Ocean Currents at the Crossroads: A Scientific and Human Odyssey

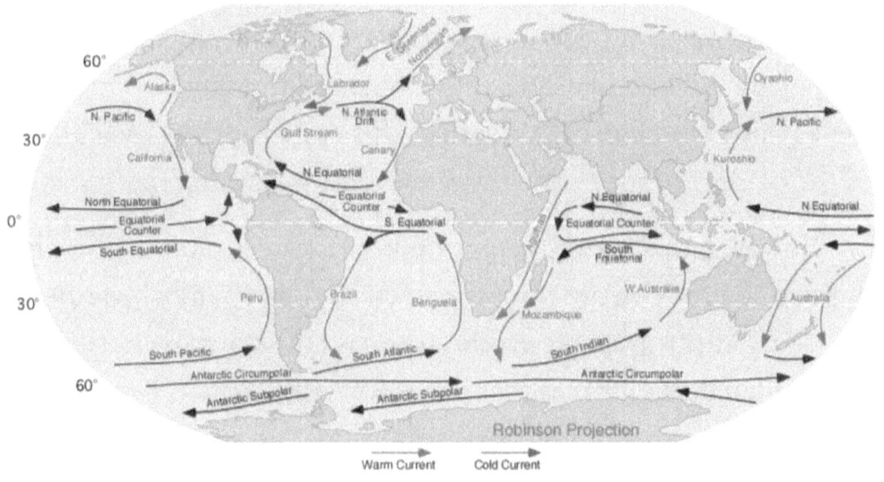

Image from Wikipedia

In the heart of every ocean, far beneath the shimmering surface, lies a vast, invisible force that has shaped the course of life on Earth for millions of years—ocean currents. These powerful water flows, driven by differences in temperature, salinity, and the winds that sweep across the planet, serve as the Earth's circulatory system. They regulate climate, transport heat, nutrients, and life across the seas, and are fundamental to the planet's ecological and climatic stability. Yet, as our world warms, these currents—once a bastion of consistency—are showing signs of faltering. The consequences of such changes are profound and could alter the trajectory of human civilization as we know it.

The intricate nature of ocean currents, often compared to the veins and arteries of a living organism, is both fascinating and critical. Just as blood circulates oxygen and nutrients to sustain life, ocean currents redistribute heat and nutrients across the globe. The world's largest and most important oceanic current system is the **Thermohaline Circulation**, often referred to as the "global conveyor belt." It connects the oceans in a continuous loop, with warm surface water flowing from the equator towards the poles and cold, dense water sinking and flowing back towards the equator in the ocean's depths. This system drives the climate of the planet, supporting ecosystems and sustaining marine life.

However, this vital system is now at risk, teetering on the edge due to the very climate it once helped stabilize. Global warming, caused by anthropogenic greenhouse gas emissions, is altering the delicate balance of forces that drive ocean currents. The question that scientists and the global community must grapple with is: What happens when these currents weaken, or worse, collapse?

The Atlantic Meridional Overturning Circulation (AMOC) and its Slowdown

At the center of this discussion is the **Atlantic Meridional Overturning Circulation (AMOC)**, a crucial part of the global conveyor belt. The AMOC plays an essential role in regulating the climate of the Northern Hemisphere, particularly in Europe and North America. Transporting warm water from the tropics to the North Atlantic moderates the climate of Europe, preventing it from becoming much colder despite its northern latitude.

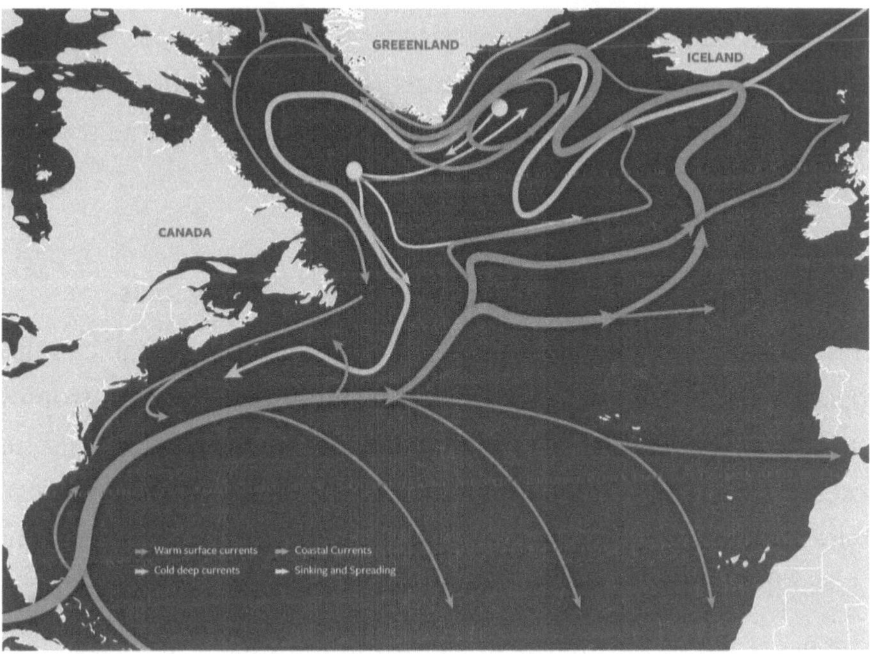

However, over recent decades, the AMOC has shown signs of weakening. Studies indicate that it has slowed by about 15% since the mid-20th century, and if current trends continue, it could decline by 34% to 45% by the end of the 21st century. The primary driver of this slowdown is the **melting of the Greenland ice sheet**, which is adding large volumes of freshwater into the North Atlantic. This influx of freshwater dilutes the salty, dense water that normally sinks and drives the circulation, reducing the force of the AMOC and leading to a cascade of climatic consequences.

The weakening of the AMOC could have far-reaching effects, including a significant cooling of Europe, more extreme weather in North America, and changes in rainfall patterns in regions as distant as the Sahel in Africa. The most alarming scenario would be a complete collapse of the AMOC, which, while considered unlikely before the year 2100, remains a possibility. If this were to occur, Europe could experience temperature drops of up to **5 to 10 degrees Celsius**, potentially leading to devastating consequences for agriculture, water resources, and overall societal stability.

Global Impacts of a Weakened Ocean Conveyor Belt

The ocean conveyor belt's role in maintaining global climate balance cannot be overstated. It transfers heat from the tropics to the poles, regulates rainfall patterns, and supports marine ecosystems. If this circulation weakens, the planet will face a future where regional climates could shift unpredictably, potentially destabilizing food systems, economies, and entire communities.

For instance, one of the most profound impacts of a weakened AMOC is the potential shift in **tropical rain belts**, which could drastically alter the **monsoon systems in Asia**. The monsoons, which are vital for agriculture and water resources in countries like India and Bangladesh, depend on stable oceanic and atmospheric circulations. A shift in these rain belts could mean more droughts in some areas and devastating floods in others, impacting millions of people and threatening food security.

Moreover, a weakened AMOC could exacerbate the frequency and intensity of extreme weather events. Warmer ocean waters, particularly along the **East Coast of the United States**, are already linked to more powerful hurricanes. Studies show that hurricanes such as **Hurricane Irene** and **Superstorm Sandy** were fueled by unusually warm waters, a phenomenon expected to worsen if ocean currents continue to slow. These storms cause billions of dollars in damage, disrupt lives, and underscore the vulnerability of coastal communities.

Another critical consequence of a disrupted thermohaline circulation is its impact on **marine ecosystems**. Ocean currents are responsible for upwelling, the process by which deep, nutrient-rich waters are brought to the surface, sustaining marine life. A weakened current system could reduce upwelling, leading to nutrient-deprived surface waters, which in turn would reduce fish populations and impact global fisheries. This would be catastrophic for the millions of people who rely on the ocean for food and livelihood, especially in developing countries.

The Role of Ocean Currents in Carbon Sequestration

The ocean also serves as a major carbon sink, absorbing around **25% of the CO2** that humans release into the atmosphere. This process is facilitated by ocean currents, which transport carbon-rich waters to the deep ocean, where it can be stored for centuries. However, a slowdown in ocean circulation would reduce the ocean's ability to absorb CO2, leading to higher levels of carbon in the atmosphere and accelerating global warming.

In a 2019 study published in *Nature*, researchers found that a weakened AMOC could reduce the ocean's capacity to store carbon, potentially leading to an additional **50 ppm (parts per million)** of CO2 in the atmosphere by the end of the century. This would further exacerbate the warming of the planet, creating a feedback loop that could push global temperatures to dangerous levels.

Scientific Debate and Uncertainty

Despite the overwhelming evidence of a weakening AMOC, there are some scientists who caution against drawing definitive conclusions. The ocean is a highly complex system, and natural variability plays a significant role in the behavior of ocean currents. Some researchers argue that the changes we are observing may be part of a long-term natural cycle, rather than a direct consequence of human-induced climate change.

However, the **Intergovernmental Panel on Climate Change (IPCC)**, which is regarded as the authoritative voice on climate science, asserts that it is "very likely" that the AMOC will continue to weaken throughout the 21st century due to global warming. Climate models that incorporate greenhouse gas emissions show a clear link between human activities and the observed changes in ocean circulation patterns. While natural variability cannot be ruled out entirely, the rate and magnitude of the changes are consistent with the rapid increase in global temperatures.

The Path Forward: What Can Be Done?

The fate of the ocean currents, and by extension, the fate of our climate, is not set in stone. There is still time to act, but the window of opportunity is closing rapidly. To protect the AMOC and other critical components of the Earth's climate system, we must address the root cause of the problem—**greenhouse gas emissions**.

According to the **IPCC**, limiting global warming to **1.5°C above pre-industrial levels** would significantly reduce the risk of a catastrophic slowdown or collapse of the AMOC. This will require a global effort to reduce emissions, transition to renewable energy sources, and adopt sustainable practices in agriculture, industry, and transportation.

Furthermore, it is essential to support and expand **scientific research** on ocean currents and their role in the climate system. While we have made significant strides in understanding the AMOC and other currents, there is still much we do not know. **Investments in ocean observation systems**, such as the RAPID array in the North Atlantic, are critical for monitoring changes in real time and providing the data needed to improve climate models.

Forests on the Frontline: A Call to Save the Planet's Lungs

As the Earth spins on its axis, vast rainforests breathe life into our atmosphere. These dense, verdant expanses—found in the Amazon Basin, the Congo, Southeast Asia, and other regions—are much more than just a collection of trees. They are intricate, living systems integral to the planet's survival. Known as the "lungs of the Earth," forests absorb vast amounts of carbon dioxide and release oxygen, enabling life to thrive. But as wildfires rage and chainsaws hum with ever-increasing intensity, our forests are being destroyed at an alarming rate. And with their destruction comes a wave of crises, from biodiversity loss to climate change acceleration.

India, with its unique position in the tropics and rich tradition of environmental stewardship, stands both as a witness and participant in this global struggle. The battle to save the world's forests is not one that can be fought in isolation. It requires coordinated action—locally and globally—driven by both science and policy and inspired by the wisdom of indigenous cultures that have long acted as the custodians of these sacred spaces.

In this chapter, we will delve into the science behind forest ecosystems, the direct impacts of deforestation, and the growing threat of wildfires, all while drawing on the latest scientific evidence. We will also examine the international and national regulations that exist to curb these threats and explore how India and other nations can step up in this critical moment. With the right action, we can still secure the future of these vital ecosystems.

Forests: The Pillars of Global Climate

Forests, particularly tropical rainforests, play a critical role in regulating the Earth's climate. They cover about **31% of the planet's land area**, but they are responsible for absorbing approximately **30% of global carbon dioxide emissions** from the atmosphere, making them one of the most powerful natural carbon sinks. According to the **Intergovernmental Panel on Climate Change (IPCC)**, deforestation and forest degradation contribute to about **13% of total global greenhouse gas emissions**, mainly due to the release of carbon stored in trees.

Each year, the world's forests absorb an estimated **2.4 billion metric tons of CO2**, significantly mitigating the impact of fossil fuel emissions. The Amazon rainforest alone stores up to **86 billion metric tons of carbon** in its trees, roots, and soils, making it a crucial buffer against the accelerating effects of climate change. However, with deforestation rates climbing—

over **10 million hectares of forest lost annually**—this natural balance is under severe threat.

The Amazon: A Tipping Point

The Amazon Basin is home to the world's largest rainforest, covering **5.5 million square kilometers** across nine countries. It is a haven for biodiversity, sheltering about **10% of the world's known species**. Yet, deforestation in the Amazon has reached record levels. According to the **Brazilian National Institute for Space Research (INPE)**, deforestation in the Amazon increased by **22% in 2021**, with over **13,000 square kilometers** of forest cleared—equivalent to the size of Northern Ireland.

This rapid loss of forest cover brings the Amazon closer to a critical tipping point. Scientists warn that if deforestation exceeds **25%**, the rainforest could shift from being a net carbon sink to a net carbon emitter. This phenomenon, known as "dieback," would lead to a self-sustaining cycle of forest loss, turning vast stretches of the Amazon into savannah-like landscapes. In turn, this would release billions of tons of CO_2 into the atmosphere, accelerating global warming and destabilizing regional climates.

Wildfires: Nature's Alarm Bells

While deforestation is often driven by human activities like agriculture and logging, another growing threat is wildfires. Historically, rainforests like the Amazon and Congo Basin have been relatively resistant to wildfires due to their humid climates. However, with rising global temperatures, these regions are experiencing an increase in the frequency and intensity of forest fires.

In 2019, the world watched in horror as wildfires engulfed large portions of the Amazon. An estimated **906,000 hectares** of forest were destroyed, releasing **228 megatons of CO2** into the atmosphere, according to data from the **Global Fire Emissions Database (GFED)**. Fires are not only a byproduct of climate change but also exacerbate it. When forests burn, they release stored carbon back into the atmosphere, creating a feedback loop that accelerates global warming.

India has not been immune to this trend. In the past decade, forest fires have increased in the country, particularly in the Himalayan region and the Western Ghats. In 2020 alone, India recorded over **37,000 forest fire incidents**, with severe impacts on biodiversity and air quality. These fires are often fueled by rising temperatures and prolonged dry seasons, both of which are linked to climate change.

Deforestation and Biodiversity Loss

The world's rainforests are not just climate regulators; they are also among the most biodiverse ecosystems on Earth. The Amazon, for instance, is home to over **400 billion individual trees** across **16,000 species**, providing habitat for thousands of animal species, including jaguars, sloths, and numerous birds and insects. However, as forests are cleared, many of these species face the threat of extinction.

The **World Wildlife Fund (WWF)** estimates that around **100,000 species** are lost every year due to deforestation, with tropical rainforests being the most affected. When a forest is cut down, entire ecosystems are destroyed, disrupting food chains and ecological balances that have evolved over millennia. A study published in the journal *Nature* found that deforestation in the Amazon is driving the extinction of species at a rate that could see up to **40% of the region's species** lost by the end of the century.

In the Congo Basin, deforestation threatens some of the world's most iconic species, including **forest elephants**, **mountain gorillas**, and **chimpanzees**. Forest elephants, in particular, play a crucial role in maintaining the ecological balance of the forest by dispersing seeds and creating clearings that allow sunlight to reach the forest floor. However, their numbers have declined by **over 60%** in the past decade due to habitat loss and poaching.

The Role of Forests in Water and Climate Regulation

Forests are not only carbon sinks; they also play a pivotal role in regulating the Earth's water cycle. Through a process known as **evapotranspiration**, trees release moisture into the atmosphere, contributing to cloud formation and rainfall. The Amazon alone produces **20% of the world's freshwater** through its river systems and rain, which feed into the atmosphere and sustain agricultural regions across South America.

Deforestation disrupts this delicate balance. As trees are removed, less moisture is released into the atmosphere, leading to reduced rainfall and prolonged droughts. This is particularly evident in regions like Southeast Asia, where deforestation for palm oil plantations has led to a significant decline in rainfall patterns, affecting both local agriculture and global food supply chains.

The Congo Basin, similarly, plays a crucial role in regulating the water cycle of Central Africa. By maintaining soil moisture and influencing regional rainfall patterns, the forest helps stabilize the climate across the continent. However, deforestation in the Congo Basin could disrupt these mechanisms, leading to shifts in precipitation that could result in more frequent droughts and water shortages in already vulnerable regions.

The Impact of Deforestation on Indigenous Communities

The human cost of deforestation is immense, particularly for indigenous communities who have lived in harmony with these forests for centuries. The Amazon is home to approximately **400 indigenous groups**, many of whom rely on the forest for food, medicine, and cultural practices. As deforestation accelerates, these communities face displacement, loss of livelihoods, and the erosion of their cultural heritage.

In India, indigenous tribes such as the **Adivasis** are similarly affected by deforestation. The Adivasis, who inhabit the dense forests of central and eastern India, rely on the forest for sustenance and spiritual practices. However, mining, logging, and infrastructure projects have encroached upon their lands, forcing them to migrate to urban areas where they often face discrimination and poverty.

International initiatives like **REDD+ (Reducing Emissions from Deforestation and Forest Degradation)** aim to provide financial incentives for forest conservation, but their success depends on the meaningful inclusion of indigenous voices. Without their active participation, efforts to protect forests are unlikely to succeed.

Global and National Regulations to Combat Deforestation

In recent years, there has been a growing recognition of the need to protect the world's forests through both global and national regulations. The **Paris Agreement**, adopted in 2015, acknowledges the importance of forests in mitigating climate change and calls for enhanced efforts to reduce deforestation. Under this framework, countries are required to include forest conservation in their Nationally Determined Contributions (NDCs).

Brazil, home to the majority of the Amazon rainforest, has committed to ending illegal deforestation by 2030 and restoring **12 million hectares** of forest land. However, political and economic pressures have made this goal difficult to achieve. In 2019, Brazil saw a surge in deforestation as the government rolled back environmental protections, sparking international outrage.

In the Congo Basin, governments have adopted the **Central African Forest Initiative (CAFI)**, which aims to reduce deforestation through sustainable land use practices. CAFI provides financial support to countries like the Democratic Republic of Congo (DRC) in exchange for commitments to reduce forest loss. However, the region continues to face challenges due to the expansion of agriculture, logging, and mining.

The forests await our response. Will we answer their call?

The Human Factor

Anthropogenic Forces and Climate Change

As our world continues its journey through space, human activity is leaving a clear mark on the climate. The increase in greenhouse gases, caused by industrialization, deforestation, and agriculture, leads to higher global temperatures by trapping heat in the atmosphere. This influence is at the core of the current climate shifts, not merely a minor aspect of the narrative.

The expansion of industries during the Industrial Revolution has significantly contributed to climate change. Factories and power plants emit carbon dioxide CO_2 into the atmosphere by burning fossil fuels such as coal and oil. The increase in global temperatures can be directly attributed to the spike in CO_2 levels. Since the Industrial Revolution began, carbon dioxide levels have increased from approximately 280 parts per million to more than 415 parts per million currently. This rise has been associated with the melting of polar ice, the escalation of ocean levels, and increasingly severe weather patterns. Forests are being rapidly deforested for agriculture and urban development despite their role in absorbing CO_2. Significant logging and farming have led to substantial deforestation in the Amazon rainforest, an important carbon sink. When trees are chopped, the carbon they hold is let out into the air, increasing the greenhouse gases that lead to global warming. The destruction of forests also interrupts the water cycle and diminishes biodiversity, leading to additional impacts on the climate.

Farming practices also contribute significantly to the production of greenhouse gases. Raising cows generates methane, which is a more potent heat-trapping gas than carbon dioxide. This is a major problem in nations with extensive cattle sectors, like Brazil and the United States. Rice cultivation, which is frequently seen in nations such as India and China, also generates methane. The application of artificial fertilizers in farming leads to nitrous oxide emissions, another powerful greenhouse gas. There are those who believe that changes in climate are a natural occurrence on Earth, using examples like the Medieval Warm Period. Nevertheless, the pace of change at present exceeds any previously observed in the geological record. Today, the composition of carbon in our atmosphere suggests that the main contributor to elevated greenhouse gases is the burning of fossil fuels rather than natural processes. This evidence indicates that human actions play a significant role in the quick transitions we are witnessing.

The impacts of these activities can be seen all around the world. Excessive CO2 absorption is causing oceans to increase in acidity, negatively impacting marine creatures such as coral reefs and shellfish. The Great Barrier Reef has undergone multiple severe bleaching episodes in recent years because of higher temperatures and rising acidity levels. Likewise, polar ice is melting at unprecedented rates, leading to increased sea levels and endangering habitats for wildlife.

Taking Responsibility

The proof indicates that human actions significantly contribute to climate change. It is important to acknowledge our responsibility in these shifts and implement measures to minimize our influence. This involves reducing emissions, conserving forests, and implementing sustainable practices. As an illustration, cities worldwide are allocating resources to renewable energy sources like solar and wind power in order to decrease their dependence on fossil fuels. Certain nations are enforcing stringent rules to decrease deforestation and support reforestation initiatives.

Moreover, sustainable farming methods like permaculture and agroforestry are being advocated to diminish methane and nitrous oxide emissions from farming. These actions not only reduce the impact of climate change but also enhance soil health and promote biodiversity. The decisions we make now will influence the future of our planet and impact the well-being of future generations. The direction ahead is evident. It is our duty to take accountability for our deeds and collaborate to build a more resilient future. We can maintain a stable and healthy climate for everyone by decreasing our carbon footprint, preserving natural ecosystems, and adopting cleaner, greener technologies.

Ethics of Environmental Stewardship

Have you ever considered how each decision you make can impact more than just what is directly around you? Every choice has the potential to impact forests, oceans, and the air we breathe. Take a moment to pause and consider it. What duty do we hold in this interlinked network of existence? Today, our impact on the environment is greater than ever before, and its importance cannot be emphasized enough. We need to view this problem as more than just a scientific or economic task but also as a significant ethical responsibility. Our duty to our planet is just as crucial as any other moral duty, but it is frequently overlooked in our daily lives. We are confronted with a difficult issue - a climate emergency that endangers the ecosystems essential to our survival. Our constant need for resources and lack of coordinated efforts are the main causes of this crisis. The central challenge in The Last Degree is to take the lead in the Climate Crucible.

Many think that solutions are found in technology, policy changes, or international agreements. Although they are crucial, they frequently overlook the underlying ethical basis necessary for significant change. Typically, symptoms are treated instead of tackling the underlying cause, similar to using a bandage on a wound that requires more extensive treatment. The true answer necessitates a change in our perception of our role in the world, not just relying on science or politics. We must view ourselves as Earth's guardians, responsible for safeguarding it for the sake of future generations. This concept is not novel - numerous indigenous societies have always valued this duty towards the environment. For instance, the Māori community in New Zealand follows kaitiakitanga, which focuses on protecting the environment on land and in the ocean. In the same way, Native American customs frequently discuss the importance of making choices while thinking about the well-being of the "seven generations" ahead. These viewpoints serve as a reminder that stewardship of the environment is not only a current issue but also a fundamental moral belief.

Why would you, with this book in hand, be concerned about environmental stewardship? Since it's not a concept detached from reality—it's closely individual. The health of the planet directly impacts the air, water, and food your family relies on for sustenance. In cities like New Delhi, India, where levels of air pollution frequently surpass safe thresholds, inhabitants, particularly children and the elderly, face serious respiratory issues. My son had difficulty breathing because of bad air quality, which was caused by the degradation of the environment. The future of a habitable world for our children depends on the actions we take now.

My own son once struggled to breathe due to poor air quality. It has hence been my mission to build Ambee and bring the power of climate data to the world for a decade now. Consider the emotional toll of witnessing a child struggle to breathe, understanding that their suffering is influenced by environmental factors. Experiencing this pain is something no parent should go through, but unfortunately, it is happening more frequently as pollution levels increase. In urban areas such as Delhi, Beijing, and Los Angeles, smog and air pollution remain ongoing issues that impact millions of individuals on a daily basis. In order to make a change, we must form emotional connections with the consequences of our environmental decisions. The proof of our influence is evident, seen in the pollution in our urban areas and the pale coral reefs in our seas. Since 1995, more than half of the coral in the Great Barrier Reef has disappeared because of increasing sea temperatures and ocean acidification. Refusing to acknowledge this truth is no longer a possibility. It is important for us to acknowledge that we are a component of an interconnected ecosystem, and our well-being is directly linked to its vitality. The impact we leave behind will not be measured by our riches or scientific achievements but by the condition of the environment we pass on.

Picture yourself standing at the edge of the sea, observing as every wave surpasses the one before it by just a bit. The increasing water levels are more than just a natural event- they are caused by human activity, indicating the melting of ice caps and the rise of sea levels. In locations such as the Maldives, the risk of rising sea levels is not something far off in the future but is a current peril that could potentially inundate the entire nation. It serves as a clear warning that climate change is not something far off in the future; it is happening at this moment, altering our world and our impact.

It's time to shift our perspective on our connection to the Earth, viewing ourselves as an integral part of a greater entity instead of dominating conquerors. We need to embrace a mindset that views us as guardians and nurturers, recognizing that every decision we make influences the fate of our surroundings. The kindness we exhibit now will have a lasting impact, demonstrating our reverence for all living beings. In my role as a co-founder and Chief Technology Officer at Ambee, I specialize in sustainability and climate change research, committing my professional life to offering practical data to inform decision-making. This involves up-to-date environmental information on air quality, temperature, and humidity that aids cities such as London and Singapore in developing improved climate action strategies. In this book, I am not merely sharing information or suggesting measures - I am urging for an ethical awakening to recognize that we all have a responsibility to protect this valuable, vulnerable planet.

The principles of environmental stewardship are not merely abstract concepts; they are crucial for our existence. As we confront the challenges of climate change, they must lead our decisions and behaviors. We have the ability to take charge of this transformation. We need to implement decisive measures and enact essential reforms in order to prevent catastrophe. Are you willing to get involved in this initiative, not just watching from the sidelines but actively taking part? Are you willing to make an effort to live in a more responsible manner and motivate others to do so as well? You have the decision and must take action immediately. We should forge a fresh trail together for the benefit of our Earth, our kids, and future generations. It is our opportunity to mold the future, and we are responsible for determining it.

Urbanization and the Environmental Footprint

Every large city has two narratives: one of human success and the other of the burden on the environment. **Urbanization: A Double-Edged Sword**

Urbanization is a global phenomenon. As of 2023, more than **56% of the world's population** lives in cities—a figure expected to rise to **68% by 2050**, according to the **United Nations**. This migration to urban areas has brought economic prosperity, better access to education and healthcare, and countless innovations. However, this concentration of people also intensifies pressure on the environment.

When cities grow unchecked, they expand horizontally into surrounding natural landscapes, causing deforestation, loss of biodiversity, and habitat destruction. **Urban sprawl**—the uncontrolled spread of urban development into rural areas—creates large, low-density suburbs that often require more energy and resources to maintain. On the other hand, vertical growth, while more space-efficient, poses its own environmental challenges, particularly related to energy consumption and waste production.

The **urban heat island (UHI)** effect is one of the most visible manifestations of these challenges. In cities like **New York**, **Tokyo**, and **Beijing**, urban temperatures can be several degrees higher than in surrounding rural areas. This is primarily due to the heat-absorbing properties of concrete, asphalt, and glass—materials that dominate urban landscapes. Studies show that **urban heat islands can increase city temperatures by 1.5°C to 3°C** on average, and during heatwaves, the impact can be even more severe. The increased energy demand for air conditioning further strains electrical grids and contributes to greenhouse gas emissions.

Energy Consumption and Carbon Footprint

Cities, despite covering only **3% of the Earth's land**, account for nearly **70% of global energy consumption** and **75% of global carbon emissions**. This disproportionate energy use is driven by the concentration of industrial activities, transportation networks, and residential energy demand. Fossil fuels, primarily coal, oil, and natural gas, remain the dominant energy sources for most cities, exacerbating the global carbon footprint.

Take **Los Angeles**, for example, a city known for its sprawling suburbs and heavy reliance on cars. Despite efforts to introduce electric vehicles and public transportation systems, the city's daily emissions from vehicular traffic remain staggering. The **average American household** emits **7.5 metric tons of CO2 annually** just from vehicle use alone. Now multiply that by the millions of cars that traverse L.A.'s highways daily, and the environmental burden becomes clear.

However, cities like **Copenhagen** are showing the world a different way forward. By investing in **wind energy** and aiming for **carbon neutrality by 2025**, Copenhagen is reducing its carbon footprint and setting an example for others. Similarly, **Vancouver** has pledged to derive **100% of its energy from renewable sources** by 2050, integrating solar power, hydroelectricity, and green building codes into its urban planning framework.

Waste Generation and Management

Urbanization generates enormous amounts of waste, with cities like **New York** producing around **12,000 tons of waste per day**. Much of this waste ends up in landfills, where it decomposes and emits **methane**—a greenhouse gas **25 times more potent than CO2**. The environmental cost of improper waste management goes beyond landfills; it contributes to pollution in rivers, oceans, and even the air we breathe.

By contrast, **San Francisco** has implemented a comprehensive zero-waste policy, diverting more than **80% of its waste from landfills** through aggressive recycling and composting initiatives. The city's success lies in its public education campaigns and strict enforcement of waste segregation at both residential and industrial levels. Similarly, **Tokyo** has pioneered waste-to-energy plants that convert the city's waste into electricity, reducing the need for landfills and cutting down on emissions.

Water and Air Quality

As urban populations grow, so does the demand for clean water. Cities such as **Cape Town** have already faced severe water shortages, with the infamous **Day Zero**—when the city came perilously close to running out of water—serving as a stark reminder of the fragility of urban water supplies. Excessive water use, pollution from industrial activities, and the destruction of natural water catchments all contribute to the growing water crisis in urban areas.

Urban air quality is another critical issue. Cities like **Beijing**, **Delhi**, and **Mexico City** have long struggled with dangerous levels of air pollution, primarily caused by vehicle emissions, industrial processes, and coal burning. **The World Health Organization (WHO)** estimates that **4.2 million people die each year** as a result of exposure to outdoor air pollution, with urban areas being the most affected.

The Promise of Sustainable Cities

While the environmental footprint of traditional cities is significant, the rise of **sustainable urban planning** offers a blueprint for a different kind of urban future—one where the growth of cities goes hand in hand with environmental preservation.

1. **Green Building Practices**: Sustainable cities prioritize the construction of **energy-efficient buildings** that minimize energy consumption and reduce carbon emissions. These buildings are often equipped with **solar panels, rainwater harvesting systems,** and **smart energy management** technologies. The use of **green roofs** and **vertical gardens,**

as seen in **Singapore**, not only reduces energy consumption but also helps cool urban areas, combat air pollution, and enhance biodiversity.

2. **Public Transportation and Mobility**: Sustainable cities reduce reliance on private vehicles by promoting **public transportation**, cycling, and walking. **Amsterdam** is a prime example of a city that has embraced sustainable mobility, with **bicycles accounting for over 60%** of all trips within the city. By investing in extensive bike lanes and integrating public transportation options, cities can reduce traffic congestion, improve air quality, and lower carbon emissions.

3. **Waste Management and Circular Economy**: In sustainable cities, waste is viewed not as a problem but as a resource. Cities like **San Francisco** have adopted the concept of a **circular economy**, where waste materials are recycled and repurposed to minimize the need for raw material extraction and reduce landfill use. By closing the loop on waste, cities can significantly reduce their environmental impact.

4. **Urban Green Spaces**: Green spaces are essential to sustainable urban living. They not only provide residents with access to nature but also help absorb CO_2, reduce the urban heat island effect, and support local biodiversity. Cities like **Melbourne** are leading the way by creating new parks, expanding tree canopies, and incorporating green corridors into urban design. Melbourne's **Urban Forest Strategy** aims to double the tree canopy by 2040, improving both air quality and the quality of life for residents.

5. **Renewable Energy Integration**: Cities can dramatically reduce their carbon footprint by investing in **renewable**

energy sources like wind, solar, and hydroelectric power. **Reykjavik** in Iceland is one of the world's greenest cities, generating nearly **100%** of its energy from renewable sources. By harnessing geothermal and hydroelectric power, Reykjavik has drastically reduced its dependence on fossil fuels and serves as a model for other cities.

Urbanization in the Age of Climate Change

As the world grapples with the realities of **climate change**, urban areas are increasingly vulnerable to its effects. Rising sea levels, more frequent and intense storms, and extreme heat pose significant risks to cities, especially those located in coastal or flood-prone areas. For example, cities like **Miami** and **Jakarta** are already experiencing the impacts of rising sea levels, with frequent flooding threatening infrastructure and displacing communities.

To mitigate these risks, cities must prioritize **climate resilience** by investing in **flood defenses**, improving **stormwater management systems**, and enhancing **disaster preparedness**. Cities like **Rotterdam** in the Netherlands are leading the way with innovative approaches to water management, including **floating buildings**, **water plazas**, and green rooftops designed to absorb excess rainfall and reduce the risk of flooding.

The Path Forward: A Global Effort

The transformation of cities into sustainable hubs requires a collective effort from governments, businesses, urban planners, and citizens. Global frameworks like the **United Nations' Sustainable Development Goals (SDGs)** and the **Paris Agreement** provide a roadmap for achieving sustainable urbanization. These frameworks call for cities to reduce their carbon footprint, improve resource efficiency, and promote inclusive, green urban spaces.

However, the responsibility does not lie solely with policymakers. Each of us has a role to play in building sustainable cities. Whether by choosing public transportation over driving, supporting green building practices, or advocating for climate-friendly policies, our individual actions contribute to the broader movement toward sustainability.

Agriculture: Feeding Us or Failing Us?

Farming has always been an essential aspect of human survival. The act of cultivating the land, sowing seeds, and reaping crops is what provides us with sustenance. However, a contradiction as complex as carbon emissions exists within these fields: the soil that nourishes billions also plays a part in climate change. In what way does agriculture, which supplies our food, also affect the Earth's temperature increase? In order to comprehend this, we must examine the positives and negatives of agriculture. Agriculture supports human life by growing crops and raising animals. Additionally, it plays a role in climate change by means of deforestation, methane emissions from livestock, and extensive use of chemicals. Around 24% of worldwide greenhouse gas emissions come from agriculture. This involves methane emitted by cows and rice fields, as well as nitrous oxide from soils with excessive fertilization. In order to address the climate emergency, it is important to analyze the methods that sustain us and discover ways to establish a more positive connection with our planet.

Observing expansive green fields and the mechanical actions of tractors clearly illustrates the shared objective of all farming techniques: to optimize the productivity of the land in order to nourish an expanding population. All farmers, whether they cultivate vast wheat fields or small organic vegetable plots, strive for a successful harvest. Yet, upon closer examination, the distinctions in farming techniques become evident. Industrial agriculture, a large part of worldwide food production, depends heavily on artificial fertilizers, pesticides, and monoculture fields with a single crop. In Punjab, India, growing wheat and rice in monocultures has caused major environmental problems. The Green Revolution, initiated in the 1960s, boosted crop production but led to soil erosion, depletion of groundwater levels, and heightened reliance on pesticides, affecting both soil conditions and water purity.

On the other hand, sustainable methods emphasize a variety of crops, organic pest control, and natural farming techniques. In India, there is a growing interest in traditional farming methods such as mixed cropping, crop rotation, and organic farming. In regions such as Sikkim, which transitioned to being entirely organic in the year 2016, farmers enhance soil fertility and increase crop durability by utilizing compost and natural fertilizers. This method decreases the necessity for chemical inputs, supports biodiversity, and improves the soil's capacity to hold water and nutrients. Can you imagine it in your mind's eye? Vast stretches of one type of plant are uniform in variety, as opposed to a varied combination of plants thriving together. The distinction extends past looks; it impacts the surroundings. Monocultures, despite being efficient, are at a greater risk of being attacked by pests and diseases, often leading to the need for more chemicals that can harm the soil and water. On the other hand, a variety of crops imitate natural environments, boosting their resilience and decreasing the necessity for toxic substances. Studies indicate that farms with diversity are better able to withstand climate-related challenges like droughts and floods because they support the health of soils and water systems. For instance, in Maharashtra, farmers who utilize organic farming methods and diversify their crops are more resilient to changing rainfall patterns due to climate change.

This comparison uncovers crucial observations. Traditional agriculture, commonly viewed as obsolete, offers important insights for promoting sustainability. These methods promote harmony that is often disturbed by industrial farming through crop rotation, soil enrichment with organic matter, and the encouragement of biodiversity. In India, techniques such as agroforestry, which involve combining trees with crops and livestock, are experiencing a resurgence, particularly in regions like Karnataka. This practice enhances soil quality, boosts biodiversity, and offers extra income opportunities to farmers. Today, the significance of sustainable agriculture is more evident than before. With the escalation of the climate crisis, the need to reduce greenhouse gas emissions and safeguard ecosystems becomes increasingly pressing. Traditional agricultural practices, previously ignored in favor of industrial productivity, now provide a route to sustainability and ecological well-being.

Can sustainable agriculture provide enough food for a global population of eight billion individuals? The solution is not simple. It requires innovation by blending modern efficiency with traditional ecological knowledge. Picture a farm utilizing advanced technology to deliver water and nutrients specifically to required areas. In India, Zero Budget Natural Farming ZBNF is promoted in Andhra Pradesh and utilizes local resources to improve soil fertility and plant health without using chemical inputs. Imagine fields adorned with cover crops that halt erosion and enhance soil health, along with natural pest control techniques that reduce the need for chemicals. This farming outlook combines traditional teachings with modern advancements.

While the sun descends over a meadow filled with wildflowers and the hum of bees, it is crucial to consider the future of agriculture. The final degree is not solely about the heat level; it signifies the slim margin in which we need to act to avoid a climate catastrophe. Will we accept the challenge of creating farming methods that support us without damaging the Earth, using a mix of old-fashioned tools and modern technology? It is up to us to make the decision. We have the option to stick with the usual methods or venture into new territory in terms of sustainability and innovation. The invitation to take action is evident - are you willing to participate in guiding the path towards a sustainable farming future that benefits both individuals and the planet?

Psychological Drivers Behind Climate Inaction and Action

It is simple to imagine melting ice caps, rising sea levels, and extreme weather when considering climate change. However, the human mind is frequently overlooked as another crucial factor. Understanding the reasons why individuals and authorities sometimes do not take action on climate change is equally crucial as comprehending the scientific aspects of the issue. The causes of not taking action are diverse and complicated, including denial, apathy, fear, and a feeling of being unable to do anything. Denial is a significant factor that contributes to a lack of action. There are individuals who deny the reality of climate change or attribute it to non-human sources. This rejection may serve as a defense mechanism against unpleasant realities. Acknowledging climate change's reality can result in anxiety and powerlessness, prompting individuals to opt for denial as a coping mechanism. In the United States, studies indicate that approximately 10% of the population continues to deny the reality of climate change or attributes it to natural environmental fluctuations despite significant scientific proof. In nations where policymakers minimize the dangers of climate change, this stance may become widespread, impacting public perception and impeding policy initiatives.

Another factor is indifference. Even if people recognize climate change, they may still feel detached from the issue. This disconnect is frequently based on the idea that climate change is a remote issue that will impact future generations or individuals in distant locations rather than their own daily lives. An example is a research conducted in Europe showing that individuals residing in coastal cities were more inclined to view climate change as a danger in comparison to those residing inland. This indicates that individuals who experience firsthand the effects of climate change are more inclined to be worried and respond, whereas those who do not experience these effects are less inclined to alter their actions.

Fear is also a major factor in the lack of action on climate change. The concept of a shifting climate may be scary, causing individuals to feel unable to act. When confronted with the enormity of the issue, individuals may believe their efforts are insignificant, prompting them to refrain from taking any action. This can be seen in communities that are already dealing with economic hardships, where climate change could appear to be another problem that they are not prepared to address. In places such as Sub-Saharan Africa, where numerous communities face poverty and inadequate infrastructure, the worry of introducing a new obstacle, such as transitioning to sustainable practices, can lead to no action being taken.

Although these mental obstacles may obstruct progress, there are also strong incentives that push individuals and authorities to tackle climate change. One of the most powerful things is consciousness. Educating individuals on the reasons and impacts of climate change increases their inclination to take action due to a greater sense of responsibility. In India, the connection between the use of fossil fuels and health issues has been highlighted by the severe air pollution in cities such as Delhi. This understanding has resulted in the implementation of strategies such as the Graded Response Action Plan GRAP, which details particular actions to address air pollution according to the air quality index. The strategy involves limiting the use of private vehicles on days with high pollution and encouraging the use of public transportation, which has proven to be somewhat effective in decreasing smog levels. The community also plays a vital role in driving action. When individuals observe their neighbors or members of their community making efforts to decrease their carbon footprint, they are more inclined to follow suit. This phenomenon is commonly known as a ripple effect, where one individual's behavior motivates others, resulting in a more widespread transformation. In Japan, community-driven efforts have effectively encouraged the adoption of sustainable practices. For instance, the town of Kamikatsu has introduced a waste management plan that urges citizens to separate their garbage into 45 different categories for recycling purposes. The collaborative work has greatly decreased the town's garbage and encouraged other Japanese regions to implement comparable methods.

In conclusion, optimism plays a key role in motivating efforts to address climate change. If individuals think their efforts can have an impact, they are more inclined to act on them. Hope can be found in concrete advancements, such as the growth of renewable energy, effective conservation initiatives, or small actions like cutting down on plastic use within a nearby area. Costa Rica's government aims to achieve carbon neutrality by 2050, sparking optimism and prompting substantial investments in green energy, forest protection, and eco-friendly farming. This idea of a sustainable future has sparked broad public backing and involvement, showcasing the influence of optimism in motivating group effort.

Both psychological factors that impede and those that drive us impact our responses to the climate crisis. With these factors, we can effectively tackle the obstacles to taking action and promote behaviors that support planet conservation. Facing climate change involves more than just cutting emissions or inventing new technology; it also requires addressing and surpassing the obstacles posed by human behavior. The ability to bring about change is found within ourselves, both in our thoughts and in the groups we belong to. It is crucial for us to acknowledge this power and utilize it in order to create a more environmentally friendly future.

The Energy Transition

Imagine a reality where the gentle sound of wind turbines and the tranquil effectiveness of solar panels supplant the loud noise of oil rigs and the bustling of coal mines. A world in which the energy used for our homes, vehicles, and industries is not sourced from fossil fuels but from natural sources surrounding us. This is not a fairy tale; it marks the start of the energy transition—a vital change from fossil fuels to renewable energy sources that is currently occurring worldwide. Energy is what has propelled human advancement across the ages. Energy has always been essential to civilization, starting from the fires used by early humans to the fuel that drove the Industrial Revolution. However, we are at a pivotal moment. The sources of energy that fueled our advancement in the past are now endangering us with a climate emergency on an unparalleled level.

The beginning of this shift in energy can be linked to the moment when humans initially harnessed natural forces for their requirements. Early forms of renewable energy included using wind to propel ships and water to power mills in ancient times. As civilizations progressed, these basic tools improved but were outshined by the finding and utilization of coal, oil, and natural gas, which appeared to provide endless energy. Progress in transitioning to modern renewable energy started to gain momentum during the 20th century. The oil crises in the 1970s served as a warning, demonstrating the dangers of depending too much on fossil fuels. During this time, there was a growing focus on alternative energies, such as wind and solar power, even though these methods were still relatively new. Every advancement in technology, starting from the initial development of solar panels to the introduction of extensive wind turbines, signifies progress along this journey.

Today, we observe a world in which renewable energy is not just a far-off vision but a tangible reality in numerous locations. For instance, in the deserts of Rajasthan, India, the Bhadla Solar Park has emerged as one of the biggest solar farms globally, utilizing solar energy to generate clean power. Due to Denmark's strong dedication to the development of renewable energy, wind power supplies almost half of the country's electricity requirements. These instances illustrate how various areas, utilizing their distinct natural assets, are playing a part in a worldwide transformation. Renewable energy is utilized in various creative ways in today's world. Solar panels are currently being incorporated into construction materials, such as solar roof tiles utilized in the US and Europe. Electric cars, which are powered by the grid, are increasing in popularity globally, with countries such as Norway seeing over 50% of new car purchases being electric. Countries like Germany are developing smart grids to manage the fluctuating nature of renewable energy sources and meet the needs of today's society, resulting in improved energy distribution efficiency.

However, transitioning to a completely renewable energy system comes with its own set of obstacles. There are major challenges to face in the realms of technology, economy, and politics. In numerous regions, the fossil fuel sector is intricately entwined with the economy, offering employment and income, thus hindering any potential changes. For example, in the United States, states that heavily depend on coal mining are encountering economic and social difficulties as they move towards cleaner energy alternatives. Yet, there is advancement happening. Germany and other countries have put in place measures to stop using coal by 2038 and are putting a lot of money into renewable energy sources. At the same time, China has emerged as the top manufacturer of solar panels globally and is swiftly increasing its wind energy capabilities. These actions are a component of a larger worldwide initiative, as seen in global pacts such as the Paris Agreement, with the goal of decreasing carbon emissions to combat global warming.

India is also a crucial participant in this shift towards clean energy. India is investing in solar, wind, and hydroelectric power with the ambitious goal of reaching 450 GW of renewable energy capacity by 2030. The government has additionally introduced programs such as the International Solar Alliance, which aims to encourage the use of solar energy in sun-rich countries. Moreover, the placement of solar panels on roofs in urban areas such as Delhi and Bangalore is increasing the availability of clean energy for residences and companies. Although these endeavors show potential, the shift to renewable energy involves more than simply switching to alternative sources of energy. A fundamental change in the way societies and economies operate is necessary. It involves blending new technologies while emphasizing saving resources and effectiveness. It is important to carefully manage a delicate balance to ensure a smooth transition without excluding anyone.

As we approach this notable transition, we need to consider: Are we willing to leave behind fossil fuels and adopt a cleaner, more sustainable future? The response needs to be a definite affirmation due to the increased consequences. Each rotation of a wind turbine, each beam of sunlight harnessed by a solar panel, represents progress towards a fresh era in human history. This chapter guides us away from the verge of climate catastrophe and towards a future where energy is utilized in sync with the Earth's natural cycles, not only consumed.

The request for action is evident. It is imperative that we take the lead in the Climate Crucible, creating a new direction where energy is a solution, not a problem. The shift to renewable energy is currently happening, and success is almost here. Would you be willing to participate in this initiative and take on a leadership role to drive it forward? It is time to take action now, as the final level of change is within our control.

Plastic Pollution and the Waste Crisis

The great Pacific Patch, 2x larger than Texas

We've all seen disturbing images: a sea turtle entangled in plastic waste, an albatross with a belly full of plastic debris, or the vast expanse of the Great Pacific Garbage Patch floating in the Pacific Ocean. These aren't isolated incidents—they're symptoms of a much larger global crisis. Plastic pollution has become one of the most pressing environmental challenges of our time. It affects ecosystems, biodiversity, and human health, demanding urgent attention and collective action.

As the co-founder and Chief Technology Officer of Ambee, I've spent much of my career analyzing and addressing environmental problems through data-driven approaches. Plastic pollution, however, is not just a problem that can be solved with technology alone—it is a systemic issue deeply rooted in our production, consumption, and waste management patterns. In this article, I'll delve into the scale of the plastic pollution problem, its far-reaching consequences, and the innovative technologies offering hope for a cleaner, more sustainable world.

The Magnitude of the Problem

Every year, the world produces over 380 million metric tons of plastic, much of which is single-use and non-recyclable. Of this, approximately 8 million tons of plastic waste enter the oceans annually—equivalent to dumping a garbage truck of plastic into the ocean every minute, according to a report by the Ellen MacArthur Foundation. By 2050, there could be more plastic in the ocean by weight than fish.

The Great Pacific Garbage Patch, an accumulation of marine debris between Hawaii and California, spans over 1.6 million square kilometers—more than twice the size of Texas. However, it's just one of many garbage patches in our oceans, and plastic pollution is not limited to these floating islands of trash.

Plastic debris has been found from the Arctic to the Antarctic, contaminating even the most remote ecosystems on Earth.

The situation is no better on land. According to a 2020 study published in the journal Science, 79% of all the plastic ever produced has accumulated in landfills or the natural environment, while only 9% has been recycled. The remaining 12% has been incinerated, contributing to air pollution and the release of toxic chemicals. This means that virtually every piece of plastic ever made still exists in some form today.

Impact on Marine and Terrestrial Ecosystems

The impact of plastic pollution on marine life is well documented. Over 700 marine species are known to be affected by plastic debris, either through ingestion, entanglement, or habitat disruption. Sea turtles, for instance, often mistake plastic bags for jellyfish, leading to blockages in their digestive systems, which can be fatal. 90% of seabirds are estimated to have ingested plastic, and the numbers are projected to reach 99% by 2050 if current trends continue.

Plastic pollution doesn't just affect marine animals—it also infiltrates the entire food chain. Microplastics, which are tiny plastic particles less than 5mm in size, are now ubiquitous in our oceans. These particles result from the breakdown of larger plastic items and are small enough to be consumed by plankton—the base of the marine food web. From plankton, microplastics move up the food chain, eventually ending up on our dinner plates. A 2019 study published in Environmental Science & Technology estimated that the average person could be ingesting up to 5 grams of plastic per week—roughly the weight of a credit card.

Microplastics have also been found in freshwater systems, soils, and even the air we breathe. In India, studies have detected microplastics in the Ganges River, one of the most iconic and vital water bodies in the world. The United Nations Environment Programme (UNEP) warns that microplastics in agricultural soils could be even more detrimental than in the oceans, as they can affect soil fertility and crop production.

Human Health Risks

The health risks associated with plastic pollution are not limited to wildlife. When plastic enters the food chain, it also poses a threat to human health. Plastics contain a variety of toxic chemicals, such as phthalates, bisphenol A (BPA), and polychlorinated biphenyls (PCBs), which can leach into the environment and accumulate in organisms. These chemicals are known to disrupt endocrine systems, cause developmental and reproductive issues, and are linked to certain cancers.

In addition to chemical exposure, the physical presence of microplastics in the human body may cause inflammation and stress at the cellular level, though more research is needed to fully understand the long-term health effects. According to a World Health Organization (WHO) report, microplastics have been detected in 83% of the world's tap water, and bottled water is not immune, with some brands containing twice the amount of microplastics as tap water.

The Role of Recycling and New Technologies

Despite the vast scale of the plastic pollution crisis, there is hope. Technological innovations and global efforts to promote recycling are paving the way for a more sustainable future. However, current recycling practices are far from adequate. Globally, only 14% of plastic packaging is collected for recycling, and the actual recycled content is even lower due to contamination and material degradation.

Advancements in recycling technologies are essential to improving these rates. Chemical recycling, for instance, has the potential to break down plastics into their original molecular components, which can then be reused to create new products. This contrasts with traditional mechanical recycling, which often downgrades the quality of plastics, limiting their use in high-quality applications.

In Japan, scientists have developed a strain of bacteria that produces enzymes capable of degrading polyethylene terephthalate (PET), a common plastic used in bottles and packaging. These enzymes can break down PET into its basic components in a matter of days, compared to the hundreds of years it takes in nature. Although still in the experimental stage, this discovery could revolutionize plastic waste management.

Bioremediation, the use of microbes to consume and degrade plastic waste, is another promising field. Researchers at the University of Portsmouth and the National Renewable Energy Laboratory (NREL) have engineered a "super enzyme" that can break down plastic bottles six times faster than previous methods. Biodegradable plastics, which decompose more quickly under certain conditions, also offer some hope, though their widespread adoption has been limited by high production costs and the lack of industrial composting facilities.

In countries like the Netherlands, initiatives such as PlasticRoad are transforming plastic waste into useful products. The PlasticRoad project repurposes recycled plastic to build roads, reducing both plastic pollution and the need for asphalt—a major contributor to greenhouse gas emissions. Early trials of this technology have been successful, and its scalability could help reduce the global plastic footprint.

While technological solutions are vital, they must be complemented by strong policy measures. Governments around the world are beginning to take action. In 2018, the European Union announced a ban on single-use plastics, which will be fully enforced by 2021, targeting products like straws, cutlery, and cotton swabs. Similar bans have been implemented in several countries, including Kenya, which now enforces one of the world's strictest plastic bag bans.

In India, the Swachh Bharat Mission (Clean India Mission) has spurred nationwide efforts to reduce plastic waste, with several states implementing bans on single-use plastics. However, enforcement remains a challenge, and there is still much work to be done to build the necessary infrastructure for waste management and recycling.

On a global scale, the United Nations Environment Assembly (UNEA) has called for a legally binding treaty to address plastic pollution. Such a treaty would hold countries accountable for reducing plastic production, improving waste management, and investing in new recycling technologies. However, progress has been slow, and the international community must act swiftly to address the growing plastic waste crisis.

The battle against plastic pollution requires a collective effort—from governments, corporations, and individuals alike. As consumers, we can take immediate steps to reduce our plastic footprint. By refusing single-use plastics, opting for reusable products, and supporting brands that prioritize sustainability, we can drive demand for more eco-friendly alternatives.

Corporations must take responsibility for their role in plastic production and waste. Many companies have already pledged to use more recycled materials in their products, but these commitments must be backed by real action. Transparency, innovation, and investment in sustainable practices are critical to creating a circular economy where waste is minimized and materials are reused. Governments must continue to implement and enforce policies that reduce plastic waste and promote recycling. Investment in infrastructure, public awareness campaigns, and international cooperation are essential to building a global solution to this crisis.

The plastic pollution crisis is not just an environmental issue—it's a reflection of our values as a society. It challenges us to rethink our relationship with the material world and recognize the interconnectedness of our actions and the planet's health. While technology offers hope, real change will only come from a shift in mindset and behavior.

As a father and an environmental advocate, I have seen firsthand the impact of pollution on our air, water, and ecosystems. But I've also seen the power of human ingenuity and the potential for positive change. We have the tools and knowledge to address this crisis—it's up to us to act. Together, we can turn the tide on plastic pollution and leave a cleaner, healthier planet for future generations.

Policy, Diplomacy, and Global Cooperation

Introduction to Global Climate Policy

The global climate policy story shows how we gradually woke up to our planet's fragility and the urgent need for everyone to act together. Initially, countries mainly focused on their own growth, ignoring the environmental impact. However, as the undeniable proof of climate change grew, we understood that it is a worldwide problem that requires a worldwide solution. The origin of international climate policy can be traced back to the late 20th century, specifically the 1972 United Nations Conference on the Human Environment in Stockholm, Sweden. This occasion symbolized an important start, uniting leaders and experts worldwide to address environmental issues for the first time. Although the discussions addressed various environmental issues, they set the foundation for raising awareness about climate change.

In the 1980s, scientists raised concerns about the rise in greenhouse gas levels. The establishment of the Intergovernmental Panel on Climate Change in 1988 marked a significant event. The IPCC evaluated climate change, its consequences, and methods to reduce its impacts. This team collected scientific evidence to inform improved policies. Their findings highlighted the impact of human activity on the climate, calling for urgent measures and paving the way for international collaboration. The 1992 Earth Summit in Rio de Janeiro signified a pivotal moment. This led to the establishment of the United Nations Framework Convention on Climate Change UNFCCC, a global pact aimed at maintaining greenhouse gas concentrations in the atmosphere. The UNFCCC recognized that countries share the responsibility of addressing climate change despite not setting mandatory emission reduction targets.

The Kyoto Protocol of 1997 was the initial significant agreement within the UNFCCC, in which developed nations agreed to legally binding emission reduction targets. This treaty acknowledged that industrialized nations, who have been the biggest contributors to greenhouse gas emissions in the air, should take the lead in combatting climate change. However, the Kyoto Protocol faced significant obstacles as well. Certain significant polluters, such as the United States, did not consent to the agreement, and developing nations were not required to make binding commitments. These problems demonstrated the difficulty of achieving consensus on climate actions. In spite of these difficulties, the Kyoto Protocol paved the way for future agreements. It implemented strategies such as carbon trading, allowing nations to reach their objectives by purchasing and selling emission permits. This method of utilizing market-driven strategies to reduce emissions continues to influence international climate policies at present. As we moved into the 21st century, the demand for climate action grew more pressing due to the escalating effects of climate change. The 2009 Copenhagen Accord aimed to create a fresh worldwide climate deal but did not meet expectations, underscoring the gap between developed and developing nations in reducing emissions. Although it had deficiencies, it represented a big leap in prioritizing climate change in global diplomatic efforts.

The global climate policy journey has seen ups and downs. Starting from basic environmental awareness to today's structured agreements, it shows a growing need for global teamwork. Each meeting, report, and deal adds to our knowledge, paving the way for stronger commitments, like the Paris Accord.

The Paris Accord: A Milestone Agreement

The Paris Accord, signed in 2015, is often praised as a major step in global climate deals. For the first time, almost every country worldwide united to promise joint action against climate change. The agreement marked a big change in how the world dealt with this tricky issue, focusing on working together, being open, and sharing responsibility.

Development and Main Objectives of the Paris Accord

The path to the Paris Accord wasn't easy. Years of talks, science warnings, and political work led to this important moment. Before Paris, global climate deals like the Kyoto Protocol had specific goals mostly for developed nations. However, these deals were criticized for not being inclusive enough and having limited reach. The Paris Accord was different. It was established due to the necessity of all nations to participate, acknowledging that climate change affects the entire planet and requires a global remedy. The aim was to cap the rise in global temperatures to less than 2 degrees Celsius from pre-industrial levels, striving to keep it below 1.5 degrees. This target was set based on scientific research that proved even a slight temperature increase could greatly affect ecosystems and human societies.

To achieve this, the Paris Accord set out several key objectives:

1. **Mitigation**: Decreasing greenhouse gas emissions in order to restrict the increase in global temperature.

2. **Adaptation**: Improving countries' capacity to adjust to climate effects.

3. **Finance**: Gathering financial resources to help developing nations with their efforts to mitigate and adapt.

The groundbreaking agreement shifted from a top-down approach, where international bodies determined targets, to a bottom-up approach. This change allowed countries to develop customized climate action strategies that are practical and tailored to each nation's circumstances.

Nationally Determined Contributions NDCs and the Flexibility of the Paris Accord

An essential aspect of the Paris Accord is the concept of Nationally Determined Contributions NDCs. Rather than being given set goals as in previous agreements, the Paris Accord required countries to determine their own methods for reducing emissions. This approach aimed to increase participation and dedication from multiple countries by allowing each to establish objectives tailored to their capabilities, resources, and economic circumstances. The NDCs' flexibility enabled countries to revise and enhance their commitments every five years. This was significant for several reasons. Initially, it acknowledged that nations are at varying levels of development and possess different abilities to tackle climate change. Additionally, it allowed countries to enhance their commitments as technology advanced and their economies evolved. This adaptability was crucial in gaining the support of major polluters such as China, India, and the United States. They were unsure about committing to strict goals before. An example is the goal set by India's NDC to increase its non-fossil fuel energy capacity to 40% by 2030, which is in line with its economic development and potential for renewable energy.

This flexibility played a key role in obtaining the support of major polluters such as China, India, and the United States. They were previously reluctant to agree to enforceable goals. For example, India's NDC set a goal to increase its non-fossil fuel energy capacity to 40% by 2030, which is in line with its economic development and renewable energy opportunities. Nevertheless, the adaptability of NDCs presented obstacles. Because individuals made their own choices about contributions, there was a wide range in the level of ambition and scope. Certain nations set ambitious targets, while others opted for more moderate objectives, raising doubts about whether the collective actions would be enough to meet the agreement's primary objectives.

Successes, Challenges, and Criticisms of the Paris Accord

Since the adoption of the Paris Accord, it has achieved numerous victories. It increased international attention and dedication to combating climate change, with more than 190 countries submitting their NDCs. The deal also sparked the creation of innovative technologies and tactics to reduce emissions. For instance, numerous countries allocate significant resources towards renewable energy, electric vehicles, and energy efficiency in pursuit of their climate objectives. Nevertheless, the Paris Agreement has encountered numerous obstacles and disapprovals. One major critique is that the agreement lacks mechanisms to ensure compliance from all parties. Countries must demonstrate their progress, yet there are no consequences for failing to achieve their objectives. As a result, there is concern that some countries may not prioritize their commitments or prioritize short-term financial gains over long-term climate goals.

Another issue arises from the disparity in the contributions and capabilities of various countries. Developing nations claim they require increased funding and technological assistance to achieve their objectives and address climate change. This has led to disputes in climate negotiations, as some emerging nations believe that wealthier countries have failed to fulfill their commitments to provide assistance.

Some critics also claim that implementing all NDCs completely would still not suffice to limit global warming to less than 2 degrees Celsius. Recent data from the IPCC indicates that the world is continuing to move towards larger temperature rises, highlighting the necessity for more ambitious measures. In spite of these obstacles, the Paris Accord continues to serve as a crucial framework for worldwide climate collaboration. It has set the foundation for future efforts to address climate change, pushing countries to enhance their pledges and work together towards a shared objective. The comprehensive scope of the agreement, along with its emphasis on transparency and accountability, has been instrumental in fostering a sense of worldwide solidarity and mutual obligation.

The success of the Paris Accord relies on countries' willingness to collaborate, assist those most impacted by climate change, and unite efforts. Progress has been achieved, however, there is still a lot to be done to achieve its objectives.

Post-Paris Developments

After the pivotal Paris Accord of 2015, the international community has been dedicated to fulfilling its obligation to tackle climate change. Significant climate conferences have taken place, where countries assess advancements, create new commitments, and address emerging obstacles. The outcomes of these meetings, along with actions by major countries, leaders, and other key players, have influenced climate policy and action in the years since Paris.

Outcomes from Major Climate Conferences

COP26, held in Glasgow in 2021, was one of the most significant climate conferences since the Paris Accord. It was a major event for worldwide climate efforts, where nations were expected to revise their pledges and showcase their advancements. The result was the Glasgow Climate Pact, aiming to speed up actions to restrict global warming to 1.5 degrees Celsius above pre-industrial levels.

At COP26, several important decisions were made:

1. **Reducing Coal Use:** The UN climate agreement made history by addressing the reduction of coal, the main contributor to greenhouse gas emissions. Countries opted to "scale back" instead of completely eliminating coal, reaching a middle ground that took into account varying energy demands and economic situations of different nations.

2. **Climate Finance:** More affluent nations reaffirmed their pledge to offer $100 billion annually to help developing countries reduce emissions and adapt to climate change. The objective for 2020 was established but not fully met, leading to new commitments and discussions on improving monetary assistance.

3. **Carbon Market Rules:** COP26 established guidelines for international carbon markets, allowing nations to trade carbon credits. It is a measure to assist nations in achieving their objectives more adaptably and increasing funding for sustainable initiatives.

Although COP26 achieved some favorable results, it received backlash for not pushing the boundaries enough. Several activists and certain developing nations contended that the pledges made were inadequate in averting the most severe consequences of climate change.

New Commitments and Revisions to NDCs

After the Paris Accord, several nations have revised and enhanced their Nationally Determined Contributions NDCs. As our knowledge of climate science expands, there is an increasing demand for more bold action to be taken. In anticipation of COP26, the European Union committed to cutting its greenhouse gas emissions by a minimum of 55% by 2030 compared to 1990 levels, an increase from its previous target of 40%. In the same way, the United States established a new goal to reduce emissions by 50-52% from 2005 levels by 2030. This marked a major change in U.S. climate policy after rejoining the Paris Accord under President Joe Biden's administration.

India, a significant producer of emissions, also has made significant pledges. During COP26, India pledged to reach net-zero emissions by 2070 and aimed to achieve 500 GW of non-fossil fuel electricity capacity by 2030. This was a significant move, given India's increasing demand for energy and dependence on coal. These revised NDCs play a critical role in advancing global initiatives to restrict temperature increases. Nevertheless, there remains a disparity between the commitments made so far and the cuts required to remain under 1.5 degrees Celsius. This has resulted in appeals for nations to raise their ambition and synchronize their policies with the most recent scientific discoveries.

The Role of Cities, Private Companies, and Non-Governmental Organizations

National governments play a critical role in global climate policy, but cities, companies, and NGOs have also become more active since the Paris Accord. They're vital for reaching climate goals. Cities are now big players in fighting climate change. They emit a lot globally but also drive innovation. An example would be Copenhagen and Amsterdam striving to achieve carbon neutrality by 2050. They're enhancing public transportation, reducing energy consumption in buildings, and supporting renewable energy. Private corporations are increasing their actions to address climate change. Microsoft and Google are working towards achieving a carbon-negative status and transitioning to using solely renewable energy sources. Government regulations, customer demand, and the conviction that sustainable practices are essential for enduring business prosperity all play a part in shaping these activities.

NGOs are essential in promoting ambitious climate actions, increasing awareness, and holding governments and companies accountable. Greenpeace, the World Wildlife Fund WWF, and groups such as Fridays for Future have played a key role in influencing public dialogue on climate change and pushing for more robust policies. The actions taken by these diverse individuals indicate a move towards a more comprehensive method for addressing climate change, as multiple groups collaborate towards shared objectives. This cooperation is essential in achieving the goals set in the Paris Agreement and ensuring a sustainable future for all.

In the future, the continued involvement of cities, businesses, NGOs, and governments will be crucial for accelerating progress towards climate objectives. Their behavior will decide our ability to address the challenge of reducing global warming and protecting our planet for generations to come.

Emerging Global Initiatives

New international efforts are arising to accelerate the transition to a more environmentally friendly future as the globe tackles climate change issues. Motivated by ambitious initiatives such as the Green New Deal, these endeavors strive to address environmental and social problems. Nations are developing carbon trading systems in order to reduce emissions. This indicates the increasing recognition that comprehensive climate action must include all individuals.

Proposals and Initiatives Inspired by the Green New Deal

The Green New Deal, originally proposed in the United States, is a comprehensive initiative aimed at addressing climate change and promoting economic and social equality. The goal is to reduce greenhouse gas emissions to net zero, create millions of high-paying jobs, and ensure clean air, water, and access to nature for all. Despite not being fully implemented in the U.S., the Green New Deal has inspired comparable proposals and discussions around the globe. In 2019, the European Union introduced the European Green Deal, which aimed to make Europe the first climate-neutral continent by 2050. The plan involves measures like investing in renewable energy, boosting energy efficiency, and supporting sustainable agriculture. It also stresses the importance of a "just transition," offering aid to workers and communities impacted by the move from fossil fuels. For instance, Poland and Spain are helping coal-dependent regions shift to greener jobs through financial aid and training programs.

In South Korea, the government introduced a "Green New Deal" as part of its wider economic recovery plan from the COVID-19 pandemic. This initiative focuses on expanding renewable energy, supporting green infrastructure, and investing in electric and hydrogen vehicle technologies. South Korea's plan also includes steps to enhance air quality, restore ecosystems, and encourage practices like recycling and waste reduction. This shows how climate action can be part of broader economic and social strategies, addressing environmental goals and economic recovery. Similarly, in 2019, New Zealand passed the "Zero Carbon Act" with the goal of reaching net zero greenhouse gas emissions by 2050. This law establishes binding targets for cutting emissions and lays out a plan for creating climate adaptation strategies. It also forms an independent Climate Change Commission to offer advice and track progress. New Zealand's approach highlights the importance of having a clear legal structure and support system to meet climate targets.

These initiatives indicate that countries are increasingly understanding the importance of tackling climate change while also supporting economic growth and social fairness. Different places may have their own policies and plans, but the core idea remains the same: creating a sustainable future involves taking comprehensive action to address environmental, economic, and social issues collectively.

International Carbon Trading Systems

Another important development in global climate policy involves the growth of international carbon trading systems. Carbon trading, also known as carbon markets, allows countries or companies to buy and sell carbon credits, representing a permit to emit a certain amount of carbon dioxide or other greenhouse gases. This market-based approach aims to offer financial incentives for reducing emissions by putting a price on carbon. The concept of carbon trading is to provide a cost-effective method to reach emission reduction targets. Countries or companies that surpass their reduction goals can sell their extra credits to those struggling to meet their targets. This system promotes investment in cleaner technologies and encourages innovation by rewarding those who reduce their emissions more efficiently.

The EU Emissions Trading System EU ETS, started in 2005, is the world's biggest carbon market. It deals with about 40% of the EU's greenhouse gas emissions, including those from power plants, industries, and airlines. The EU ETS is known for cutting emissions in the region and pushing for cleaner energy sources. However, it has had challenges like fluctuating carbon prices and doubts about its long-term impact. Lately, more places have set up their carbon trading systems. China, the top greenhouse gas emitter globally, began its national carbon market in 2021. It initially covers the power sector and will expand to industries like steel and cement later. China's move is a big deal for global climate efforts, showing its aim to cut emissions while growing its economy.

In addition to national and regional systems, efforts are underway to create international carbon trading mechanisms within the Paris Accord framework. Article 6 of the Paris Accord allows for a global carbon market, enabling countries to trade carbon credits across borders. This system aims to boost collaboration and spur more ambitious climate action by enabling countries to team up to reach their goals. While carbon trading systems offer a promising way to cut emissions, they have their critics. Some argue that carbon markets could create loopholes allowing countries or companies to dodge genuine emission cuts. Others are concerned about potential market manipulation or uneven sharing of benefits and burdens. To address these worries, it's crucial to ensure that carbon trading systems are transparent, accountable, and focused on delivering real emission reductions.

The development of new initiatives and carbon trading systems shows that achieving climate goals needs a mix of approaches. These efforts are still growing but are a key step towards more inclusive climate action. As countries keep exploring and using these strategies, their experiences will offer valuable lessons for the global community in the fight against climate change.

Challenges in Global Climate Diplomacy

To tackle climate change, countries must work together. However, global politics and conflicts can make this tough. Varying national interests, economic goals, and past responsibilities create challenges. Discussions on climate fairness, focusing on how policies affect different nations, further complicate global cooperation on climate.

Influence of International Relations and Geopolitical Conflicts

International relations are key in shaping global climate action. Countries' willingness to work together on climate issues often depends on their overall interests and relationships. For example, tensions between major powers like the United States and China can affect their collaboration on climate efforts. Both are top greenhouse gas emitters, and their teamwork is vital for global climate progress. Yet, trade disputes, global influence competition, and political differences can hinder finding common ground. Geopolitical conflicts can affect climate diplomacy, too. Take the war in Ukraine, for instance, impacting energy policies in Europe and beyond. European countries are working to lessen reliance on Russian gas, leading to a fresh focus on renewable energy. However, the urgency for energy security has led to more coal and fossil fuels use in some cases, complicating emission reduction efforts. This shows how challenging it is to balance short-term politics with long-term climate goals.

Also, local disputes can influence climate action. In the Middle East, for instance, water scarcity worsened by climate change has heightened tensions between countries sharing water resources.

Working together on climate matters in such areas often means dealing with old political and territorial disputes, which is a big obstacle.

Debates Around Climate Justice

Another big challenge in global climate diplomacy is the debate about climate justice. Climate justice means that those who have contributed the least to climate change, often developing nations, are the most affected by its impacts. This idea raises important questions about fairness and responsibility in international climate policy. Developed countries have historically contributed the most to greenhouse gas emissions and have benefited economically from industrialization. In contrast, developing countries, which are often more vulnerable to climate impacts like extreme weather and rising sea levels, have contributed much less to the problem. This difference has led to calls for developed nations to take more responsibility in addressing climate change, both by reducing their own emissions and by offering financial and technical assistance to developing countries.

These debates often come up in global climate talks. For example, at the COP26 conference in Glasgow, developing countries asked for more money from wealthier nations to handle climate change impacts and shift to cleaner energy. The $100 billion annual goal set by developed countries to aid developing nations, pledged for 2020, hasn't been fully met yet, causing frustration and distrust. Climate fairness also includes talks on "loss and damage," which refers to the unfixable impacts of climate change, such as losing land to rising seas or livelihoods to extreme weather. Developing nations say they should be compensated for these losses, but many developed countries are wary of committing firmly and worried about the costs.

These debates show the challenge of balancing economic growth with environmental sustainability. Developing countries focus on economic progress and reducing poverty, which can clash with strict climate policies. For instance, India aims to lift many out of poverty while working on emission cuts and boosting renewable energy. This balancing act between development and climate duties is a key issue in global climate talks. Despite the hurdles, there are signs of progress. Some developed nations have upped their financial support for developing countries, and there's a growing understanding of the need for fair climate policies. The discussion on climate justice is slowly changing, with more focus on shared responsibilities and mutual gains.

The road to effective global climate action is complicated, with geopolitical conflicts, different national interests, and debates on fairness and responsibility. Tackling these challenges needs open conversation, mutual understanding, and a readiness to find common ground. As climate change's impacts worsen, the call for efficient and fair global cooperation grows more urgent.

Mechanisms for Enforcing Global Climate Commitments

Making sure countries stick to their climate promises is a key part of global climate policy. To reach the goals set in agreements like the Paris Accord, the world has come up with different ways to track progress and push for compliance. These methods aim to keep things clear, hold countries accountable, and inspire them to meet or even beat their targets.

Systems for Monitoring Progress

To track progress on climate commitments, we check how well countries are keeping their promises to cut greenhouse gas emissions, deal with climate impacts, and assist developing nations. The UNFCCC has set up ways to ensure that countries report their efforts accurately and on time. A key tool for monitoring is the Global Stocktake, part of the Paris Accord. It happens every five years and reviews how all countries are doing toward the agreement's long-term goals. This review looks at national reports on cutting emissions, adapting to climate change, and providing financial support. By going through these reports, the process spots gaps between what's being done now and what's needed to keep global warming under 2 degrees Celsius.

Countries must regularly submit National Reports detailing their progress towards meeting their Nationally Determined Contributions NDCs. These reports share data on emissions, policies, and the effectiveness of strategies. Technical experts review them to check for accuracy and completeness, highlighting areas needing more action. The Enhanced Transparency Framework ETF is a key part of the Paris Accord. It gives guidelines on reporting greenhouse gas inventories and NDC progress for better comparability and accountability. Developing nations get support to report accurately, ensuring all countries can fully engage in global climate efforts.

Encouraging Compliance through Sanctions and Incentives

While monitoring systems provide a way to track progress, encouraging compliance with climate commitments requires additional measures. The international community has developed a mix of approaches, including incentives for countries that exceed their targets and potential consequences for those that fall short.

Incentives for Exceeding Targets

Countries that exceed their climate promises can benefit in many ways. They might get global recognition for leading the way, boosting their position in worldwide talks and drawing investment in eco-friendly tech. For instance, nations making big strides in cutting emissions or using renewable energy could secure funds from global climate finance programs like the Green Climate Fund GCF. This money backs more climate action and helps nations keep progressing.

Sanctions for Non-Compliance

The Paris Accord doesn't have strict penalties for countries that miss their NDCs, but there are ways to push for compliance. One method is using diplomatic pressure and "naming and shaming," where countries failing to meet their commitments get called out publicly. This kind of accountability can work well, as no country wants to be seen as not doing its part in fighting climate change.

Moreover, there's talk about adding stricter penalties for not following the rules. Some ideas include tying climate progress to trade deals or financial help. For example, the European Union is thinking about putting a carbon tax on imports from countries with weaker climate rules. This move is meant to level the playing field for countries working hard to lower emissions and push others to do more, too. These approaches work best when countries are willing to cooperate and take shared responsibility.

While making sure everyone follows the rules can be tough, a mix of openness, rewards, and diplomatic pressure sets up a system that pushes countries to reach their climate targets.

As climate change worsens, enforcing rules becomes more crucial. Monitoring and encouraging compliance are vital to ensure countries act on their promises. The global community must strengthen these measures to keep climate efforts on track and hold all nations accountable.

Voices of Change

We are at a crucial moment in the fight against climate change. The discussion has moved beyond just arguments and doubts; the evidence is clear, and the urgency is real. We are seeing more than just a growing awareness; there is a strong push for real action. This energy isn't just happening in high-level meetings or universities; it's taking place in communities, businesses, farms, and especially among young people. They are determined not to accept a future marked by inaction and environmental harm.

Climate change, once considered the concern of only policymakers and scientists, is now being tackled by a wide range of people—from governments making international deals to local innovators creating new solutions, from grassroots activists fighting for environmental justice to investors shifting their money towards sustainable projects. This variety of voices and actions is creating a new approach in our shared battle against climate change.

As we watch these efforts unfold from different places and are driven by connected goals, a clearer picture starts to form. It's a patchwork of change that shows us that while the challenge is immense, the opportunity for a powerful response is even bigger.

Expanding Global Agreements

International agreements have always played a key role in the fight against climate change. They help set a plan and provide a guide for countries to follow. The Paris Accord, as discussed earlier, was a big step. It brought almost every country together with a common goal: to keep global warming below 2 degrees Celsius, ideally under 1.5 degrees. But the Paris Accord wasn't perfect. It relied on countries setting their own targets, which led to a wide range of ambitions and commitments.

Recognizing this, the Glasgow Climate Pact, agreed upon at COP26 in 2021, aimed to build on the Paris Accord by pushing for more concrete actions. One key development was the commitment to reduce the use of coal, the first time a global agreement directly addressed fossil fuel consumption. Still, the language shifted from "phase-out" to "phase-down" in the final hours of negotiations, revealing the difficulty of reaching consensus among countries with different needs and dependencies.

Another significant focus of the Glasgow Pact was climate finance. Wealthier nations promised $100 billion a year to help developing countries adapt to climate impacts and transition to clean energy. However, critics argue that these promises often fall short and that more concrete, reliable funding mechanisms are needed. The pact also called for more transparency, urging countries to provide regular updates on their climate goals to keep track of global progress and identify where urgent action is needed.

Looking ahead, the upcoming COP28 and COP29 meetings aim to address unresolved issues. A major focus will be on the "global stocktake" process, which will assess how well countries are meeting the goals of the Paris Agreement. This process is crucial for identifying gaps and ensuring that commitments turn into real actions. There's also growing recognition of the role of non-state actors—like cities, companies, and non-profits—in driving climate action. Campaigns like "Race to Zero," where cities and businesses commit to net-zero emissions, show that climate action isn't just the responsibility of national governments. These groups often bring fresh ideas and urgency that can push traditional approaches forward.

Recent policy updates in countries like India and Brazil reflect this shift. India has set ambitious targets to reach 500 GW of non-fossil fuel energy capacity by 2030 and achieve net-zero emissions by 2070. Brazil, facing criticism for its deforestation rates, has committed to cutting its greenhouse gas emissions by 50% by 2030. These are not just numbers on paper; they signal a deeper understanding that economic stability and growth depend on taking bold climate action now.

These global agreements aren't just static documents but evolving commitments that need constant revisiting, revision, and reinforcement. The real test lies in the follow-through—in how countries, cities, companies, and communities turn these promises into action. And that is where the voices of change come in, pushing, prodding, and ensuring that these commitments do not remain just words on paper.

Renewable Pioneers: Spotlight on Innovation

Around the world, pioneers in renewable energy are showing us what's possible. These aren't just theoretical projects; they're real, on-the-ground efforts that are setting new standards for clean energy.

When most people think about renewable energy, solar panels, and wind turbines usually come to mind. But there's a lot more happening. Countries like Iceland, New Zealand, and Kenya are tapping into geothermal energy, while the UK, Australia, and Canada are exploring the vast potential of ocean energy. These projects are more than just innovations; they are proving that renewable energy is both practical and powerful.

Geothermal Energy Advancements

One of the most exciting areas of renewable energy right now is geothermal energy. Geothermal energy might not grab headlines like solar or wind, but it has incredible potential. It uses the Earth's natural heat to provide a steady, reliable source of power. Almost all of Iceland's electricity and heating come from renewable sources, with geothermal playing a big role. The country's geothermal plants are so efficient and effective that they serve as models for the rest of the world. Tourists even visit them to see how they work—turning clean energy into an attraction in itself.

New Zealand is another leader in geothermal energy. For decades, the country has leveraged its unique geological features to generate electricity from geothermal sources. Projects like the Ngatamariki Geothermal Power Station are setting records for efficiency and output, showing how a focused approach to renewable energy can yield great results. In Kenya, the Olkaria Geothermal Plant is doing something similar but with a different twist. It's not just providing clean energy; it's also creating jobs and supporting local communities. For a country that has faced many economic challenges, this kind of project shows how renewable energy can drive development and provide stability.

Ocean Energy Developments

While geothermal energy is growing beneath our feet, another exciting frontier is unfolding in our oceans. The power of the sea—through tides and waves—is being harnessed in ways that were unimaginable just a few decades ago. The United Kingdom, with its long coastlines, is a leader in this space. The MeyGen project in Scotland is the world's largest tidal stream project and is setting a global standard. Using underwater turbines, it captures the natural flow of ocean tides to generate electricity. This technology is still in its early stages but is already proving to be a reliable source of clean energy that doesn't depend on the sun shining or the wind blowing.

Australia is also making strides with ocean energy. Projects like Wave Swell Energy's UniWave200 are being tested in Tasmania, using the natural movement of waves to generate power. This project is unique because it doesn't just capture energy; it's designed to integrate with existing grids seamlessly, showing how new technologies can be adapted to current infrastructures without massive overhauls.

In Canada, the Bay of Fundy, known for having the highest tides in the world, is a prime location for tidal energy projects. Companies like Nova Innovation are installing tidal turbines that harness this power to generate electricity.

While these projects show what's possible at a local level, companies are also stepping up with bold strategies that are changing the game on a global scale. One of the most notable examples is Ørsted, a company that transformed itself from a coal-heavy utility into the world's largest offshore wind developer. This kind of corporate shift isn't easy; it requires vision, commitment, and a willingness to take risks. Ørsted's journey shows how a company can align its business with the needs of the planet and thrive in the process. It's a story of reinvention, proving that big companies can change direction and lead in sustainability.

Similarly, companies like Tesla and Vestas Wind Systems are pushing the boundaries of what we think is possible with renewable energy. Tesla, known for its electric vehicles, is also a leader in battery storage, which is essential for making renewable energy reliable. Without efficient storage, much of the energy generated by renewables could go to waste. Vestas Wind Systems, on the other hand, continues to innovate in wind turbine technology, driving down costs and increasing efficiency, which makes wind energy more accessible to countries around the world.

What connects all these projects and companies is a framework for innovation built on a few key elements: solid policy support, technological advancements, and effective collaboration between the public and private sectors. Policy incentives, like subsidies for clean energy or penalties for carbon emissions, create the right environment for renewable energy to grow. Technological innovation, such as advanced battery storage or more efficient turbines, makes renewables more viable and competitive. Public-private partnerships, where governments and businesses work together, help scale these innovations quickly. These frameworks must be adjusted to fit local contexts. The key is to be flexible and learn from what works in other places, adapting successful models to meet local needs.

Innovation is shaping the future of energy, but we must also focus on taking care of the natural world that supports us. Worldwide, people are working to restore ecosystems and protect the biodiversity that keeps our planet balanced. This effort goes beyond just saving what remains; it's about bringing back what we've lost.

In Europe, projects called rewilding are gaining momentum as a way to restore nature. For instance, in Scotland's Cairngorms National Park, initiatives are underway to reintroduce native trees and animals like the Eurasian lynx. The aim is not just to plant trees or release animals but to allow nature to recover on its own. In these areas, rivers are allowed to change their courses, and wildlife plays a role in managing the ecosystem, demonstrating how a hands-off approach can be effective.

Similarly, in Asia, countries like Kazakhstan are working to bring back species such as the saiga antelope to their grasslands. These efforts show that even in places affected by human activity, there is hope for recovery if we give nature the chance to heal itself.

Indigenous-Led Conservation Strategies

While rewilding is one approach, another powerful method is led by indigenous communities. In the Amazon, groups like the Yawanawá are not only protecting their lands but also bringing together old wisdom with new technology. They see the forest not just as a resource but as a living system. Their approach is not about fencing off nature but living with it, creating a balance that benefits both people and the environment.

In the Arctic, indigenous communities like the Inuit are also at the forefront of conservation. They are using their deep knowledge of the land to protect their homes and way of life, blending traditional practices with modern science to address threats like climate change.

Maria's Rainforest Guardians

One example of grassroots action is "Maria's Rainforest Guardians." This started with a few locals who wanted to save their forest from logging. They organized, educated their community, and created alternative livelihoods that depended on keeping the forest intact. What began as a small effort grew into a national model for community-led conservation, showing how local action can lead to big changes. Today, Maria's Rainforest Guardians are not just protecting trees; they are changing attitudes, empowering communities, and influencing policies on a national scale.

Other stories echo this same spirit of determination and creativity. In Indonesia, local communities have established community-based marine protected areas to restore coral reefs and fish populations devastated by destructive fishing practices. In Canada, the Haida Nation's efforts to manage the Gwaii Haanas National Park Reserve have become a global example of indigenous-led stewardship, balancing ecological health with cultural preservation. These stories remind us that conservation isn't just about saving what's left; it's also about reclaiming what's been lost and building a future where both people and nature can coexist.

These grassroots movements are proving that change doesn't always have to come from the top. Sometimes, the most effective solutions begin in our own backyards, in our neighborhoods, and in our cities. And when these efforts gain momentum, they have the power to shape national policies and even influence global strategies.

Grassroots Movements: Local to National Influence

Grassroots movements are about people coming together to solve problems in their own communities. What makes these movements powerful is that they are driven by those who know their local context best. They see the issues firsthand and often have a clearer sense of what solutions will work. As these movements grow, they start to have an impact beyond their immediate area. They inspire others, draw attention to critical issues, and even push governments to adopt new policies.

Take urban sustainability projects as an example. In cities like New York, where space is limited, communities are turning rooftops and vacant lots into gardens. These urban gardens are more than just green spaces; they are hubs of local food production, community engagement, and education. Tokyo has also embraced this idea, with rooftop gardens helping to reduce the urban heat effect and provide a bit of green in the concrete jungle. In Singapore, the Green Mark Scheme encourages sustainable building practices, rewarding those who incorporate green roofs, efficient energy use, and water-saving measures. These initiatives might seem small on their own, but they create a culture of sustainability that can influence city planning and policies.

Agricultural Cooperatives and Sustainable Farming

Outside of cities, similar grassroots efforts are transforming rural areas. In Africa, the concept of Farmer-Managed Natural

Regeneration FMNR is changing how communities manage their land. FMNR involves local farmers in the process of naturally regenerating trees on agricultural land, which helps to restore soil fertility, retain water, and boost crop yields. What started as a small initiative in Niger has now spread across several countries, showing how local practices can lead to broader regional policies on sustainable agriculture.

In India, agricultural cooperatives are empowering farmers by promoting sustainable farming practices. These cooperatives help farmers access resources, share knowledge, and sell their products at fair prices. By working together, farmers can advocate for policies that support sustainable farming, such as subsidies for organic fertilizers or incentives for water-saving technologies. The cooperative model has become a way for farmers to take control of their livelihoods while also promoting sustainable land use.

For grassroots movements to grow and have a lasting impact, they need support. One of the key strategies is community education. When people understand the importance of sustainability and know how to implement it, they are more likely to take action. This could mean workshops on composting, classes on sustainable farming techniques, or seminars on renewable energy solutions. The more people know, the more they can do.

Access to sustainable technology is another critical piece. From solar panels to water-efficient irrigation systems, providing communities with the tools they need to implement sustainable practices can make a big difference. Often, these technologies are not accessible due to cost or lack of awareness. Finding ways to lower these barriers, whether through subsidies, grants, or community funding initiatives, can empower more people to join in.

Finally, participatory governance is essential. This means involving local communities in decision-making processes that affect their environment. When people have a say in what happens in their own neighborhoods or fields, they are more invested in the outcome. It builds a sense of ownership and responsibility, which is crucial for the long-term success of any grassroots effort. The call for sustainability is being built from the ground up, one garden, one farm, and one community at a time. And as these local efforts continue to expand, they have the power to reshape how we think about and act on climate issues at every level.

As the reality of climate change becomes clearer, the flow of money is starting to change direction. Investors, from small-scale stakeholders to major financial institutions, are beginning to realize that the old ways of investing aren't just risky for the planet—they're risky for portfolios, too. This realization is pushing a significant shift in investment strategies, leading to a rethinking of what it means to invest in our future.

The Investment Shift: Transition in Major Investment Funds

For decades, fossil fuels were seen as a safe bet for investors. Coal, oil, and gas were considered the backbone of a strong portfolio, providing steady returns year after year. But that's changing. There's a growing trend of divestment from fossil fuels, with many funds pulling their money out of these industries and looking for cleaner, more sustainable alternatives. This shift isn't just a trend; it's becoming a fundamental change in how we think about investing.

Several major funds and financial institutions have made headlines by announcing their plans to divest from fossil fuels. The Norwegian Government Pension Fund, the world's largest sovereign wealth fund, has begun to phase out investments in companies heavily involved in coal mining. Universities, like the University of California, have also pledged to divest their endowments from fossil fuels, citing both ethical reasons and the long-term financial risks associated with climate change. This wave of divestment sends a powerful message: the future of finance is green, and the days of relying on fossil fuels are numbered.

Green Venture Capital

As traditional investments shift away from fossil fuels, there's a rising interest in green venture capital. These are funds specifically targeting green innovations, focusing on technologies and businesses that aim to solve environmental problems. One of the most well-known examples is Breakthrough Energy Ventures, founded by a group of private investors, including Bill Gates. This fund invests in companies developing solutions across energy, agriculture, and other sectors to reduce greenhouse gas emissions and promote sustainability.

Impact investing funds are also gaining ground. Unlike traditional venture capital that focuses primarily on financial returns, these funds are also interested in positive social and environmental outcomes. The idea is simple: by putting money into businesses and technologies that benefit the planet, investors can earn returns while also supporting a more sustainable future. This dual focus on profit and purpose is redefining what it means to invest wisely.

Impact Investing: Emphasis on ESG Criteria

Another important part of this shift is the growing emphasis on Environmental, Social, and Governance ESG criteria in investment portfolios. More and more investors are looking at ESG factors as critical indicators of a company's long-term viability. Companies that score well on ESG are often seen as less risky and more likely to succeed in a world increasingly focused on sustainability. As a result, many large asset managers, like BlackRock and Vanguard, are now incorporating ESG criteria into their investment decisions. This shift is encouraging more companies to improve their environmental practices, social responsibility, and governance structures to attract investment.

Green Bonds and Sustainable Infrastructure

A practical example of how this shift in investment is making a difference can be seen in the growth of green bonds. Green bonds are financial instruments specifically designed to fund projects that have positive environmental impacts. Cities like Paris and New York have issued green bonds to finance everything from energy-efficient buildings to expanded public transportation systems. These bonds not only help cities reduce their carbon footprints but also offer investors a stable return on their investment.

One notable success story is the issuance of green bonds by the World Bank to fund sustainable infrastructure projects in developing countries. These projects have included everything from solar energy plants in India to wind farms in South Africa. The economic returns have been strong, and the environmental benefits are significant, showing that sustainable investments can be both profitable and impactful.

Building a Unified Voice for Change

As global agreements like the Paris Accord and the Glasgow Climate Pact take shape, the world is beginning to see a shift. Countries are being urged to meet climate targets, but it's not just about what happens at international conferences. On the ground, we're seeing companies stepping up, shifting their focus toward renewable energy. Many are starting to see it as more than just a requirement—it's a chance to lead. In parallel, conservation efforts—whether massive national projects or smaller, local initiatives—are showing us that restoration is possible, even in the face of overwhelming odds. What's perhaps most striking is the growth of grassroots movements. They remind us that change often starts within communities and can spark something bigger, even influencing policies on a larger scale.

And now, we're witnessing a significant shift in finance as well. Investors, who have been tied to fossil fuels for so long, are rethinking their strategies. They're realizing that the future of business is tied to the future of the planet, and they're adjusting accordingly. We're seeing a realignment of investment strategies, with money flowing into sustainable projects in ways we might not have predicted a decade ago.

These efforts, diverse as they are, are connected. One movement inspires the next. A new policy might spark innovation in a company, while a community project might shift investor attitudes. It's all part of a larger, interwoven effort, creating more than the sum of its parts. But even with all this progress, there's no denying that a lot of work lies ahead. The climate crisis is not a problem for someone else to solve—it's something we all have a part in.

It's easy to feel small when facing such an enormous challenge, but whether you're pushing for stronger policies, inventing cleaner technologies, directing investments toward greener options, or simply encouraging sustainable practices in your local community, your role matters. No action is too small. Every effort, no matter where it comes from, helps build the momentum we need.

The journey to a sustainable future doesn't happen all at once. It happens in the choices we make today. We all have a role to play in staying informed, involved, and proactive. And when we come together—each of us doing what we can, where we are— that's when real, lasting change happens.

Turning the Tide

Once upon a time, the future seemed like something from a science fiction movie. Robots were cleaning our homes and cities under big domes and cars that could fly. People imagined a world that was smooth, sleek, and effortless. If we could go back 100 years, the ideas about how life in the 21st century would look were limitless. But not many thought about the environmental price we would pay for chasing progress so hard.

The world we live in today isn't what we dreamed of. Instead of flying cars, we're facing extreme heat waves. Instead of robots helping us with chores, we're battling wildfires and floods. The future isn't what we expected, and a big part of that is how we've treated our planet.

The World We Might Inherit

Think about where we're headed right now. We've seen the reports, heard the warnings, and noticed the changes happening all around us. The truth is, if we keep following the same climate policies, the Earth we leave behind will be different and could be unrecognizable. The signs are already here. Every year, we experience more extreme storms, longer droughts, and unpredictable weather becoming part of our daily lives. This isn't some distant future; it's happening today, and the time to make a change is running out.

Coastal cities that we perceive as permanent fixtures, like Miami, Mumbai, and New York, are seriously threatened by increasing sea levels. Whole communities may soon have to abandon the only homes they have ever known in places like Bangladesh and the Maldives. The wildfires in Australia, the heatwaves sweeping across Europe, and the floods that devastate cities—they aren't anomalies anymore. They're becoming what we expect, year after year.

But what does the future really look like if we continue down this path? The science is clear, and the result depends on how quickly we respond. The Intergovernmental Panel on Climate Change (IPCC) has been publishing reports for years, showing us where we're headed if we don't change course. They offer several scenarios, ranging from the best case—where we keep global temperatures from rising more than 1.5°C above pre-industrial levels—to the worst, where temperatures could rise by more than 3°C. These numbers may seem small, but the consequences are not.

In the best-case scenario, where we limit warming to 1.5°C, we'll still face major challenges. Even at this level, sea levels will continue to rise, but the impact might be more manageable. Coastal cities might be able to adapt with the right infrastructure, though some low-lying areas will likely be lost. Food security could remain stable but with regional disruptions. Biodiversity will suffer, but the damage might be reversible with significant conservation efforts.

Next is the mid-case scenario, where global temperatures increase by about 2°C. This would lead to serious consequences. The polar ice sheets would melt more quickly, causing sea levels to rise significantly. Many people might have to leave their homes due to flooding, droughts, or lack of food. Ecosystems that support millions of species could fail, leading to greater food insecurity, especially in areas already struggling with drought, like Sub-Saharan Africa.

Then there's the worst-case scenario—if temperatures rise more than 3°C. In the future, the problems we're facing today will become overwhelming. Cities like New York and Mumbai could be flooded, forcing millions to move. South Asia would experience more frequent cyclones, and droughts in places like Sub-Saharan Africa could last for many years. Wildfires would hit areas like California and Australia every year, turning green landscapes into barren land. The biodiversity we depend on for clean air and food could be pushed to the brink, and entire ecosystems might fall apart.

The world we might inherit could look very different depending on the choices we make today. These projections aren't distant possibilities—they're grounded in data, and the impacts will vary across the globe. No place on Earth will be untouched, and some will feel the effects sooner than others.

These effects won't merely influence abstract figures or far-off places; they hold significance for each of us. How will your community cope with rising sea levels? Will your region face longer droughts, or will there be more intense storms? The urgency for action is not only global—it's deeply personal.

Take the Maldives, for instance. This small island nation highlights the impact of climate change. Each year, rising sea levels pose a serious threat to its existence. For the people of the Maldives, losing their homes is not some distant concern—it's a genuine fear. They are directly experiencing the reality of a future where their entire country could be underwater.

The World We Could Build

Now, let's step away from the doom and gloom for a moment. What if we actually got this right? What if, instead of chasing short-term gains, we started building a world where our cities and towns worked with nature, not against it?

Take Singapore, for example. In addition to being one of the planet's most populous areas, it leads the way in urban sustainability. Singapore's Green Plan is turning the city into a model of how urban life can be transformed. A place where the streets are bustling with electric buses and bikes instead of congested with gas-guzzling cars and where buildings are vertical gardens rather than plain concrete towers. They are promoting recycling and reducing the use of single-use plastics in an effort to create a zero-waste society. It's not a utopian dream—it's already happening, and it shows how even the busiest cities can adapt.

Denmark's energy islands are an exciting new idea that's coming to life. These islands serve as centers for renewable energy, collecting and storing wind energy from the North Sea to supply power to Denmark and nearby areas in Europe. This project shows how countries can work together to create more reliable and cleaner energy systems. It's a great example of international teamwork in the energy field, focusing on shared benefits rather than individual interests.

It goes beyond cities, though. Rural areas can lead the way, too. Entire villages in many parts of India are switching to organic farming. Instead of using dangerous chemicals, they are instead adopting natural, regenerative techniques. Not only is this beneficial to the environment, but it also generates employment, maintains healthy soil, and even brings ecotourism to certain areas. Brazil is doing something similar with agroforestry, blending agriculture with forest management. Farmers are growing food while restoring the forest, creating a system that's good for people and the planet.

These examples aren't isolated projects—they're pieces of a larger vision. Creating a sustainable world requires a solid plan that both cities and rural areas can adapt to meet their specific needs. The United Nations' Sustainable Development Goals (SDGs) provide a useful framework with clear targets for clean energy, sustainable cities, and responsible consumption. However, having a framework alone isn't enough. This is where the idea of 'Doughnut Economics' comes into play. It's a way of thinking about the growth that keeps us within the planet's limits. Instead of chasing endless growth, this model illustrates how we can thrive within the Earth's ecological boundaries. In simple terms, it's not about bigger, it's about better.

Now, picture a place like Masdar City in the UAE. It's a city, but it doesn't look like the ones you're used to. Solar panels gleam in the desert sun, providing energy to every building. The city is carbon-neutral, meaning it doesn't add to the planet's warming. People move around in electric vehicles, and the buildings are designed to use as little energy as possible. Or think about Mexico's Smart Forest City, a place where technology and nature blend seamlessly. Forests and farms grow alongside homes, and the city's systems are built to work sustainably from the ground up.

These places might sound futuristic, but they're already taking shape. They're showing us what's possible when we decide to make the future something to look forward to.

Innovative Horizons

As we move forward, it's clear that innovation will play a major role in addressing climate change. The future we want to create isn't only about changing our habits; it's about using technology to change the game. Some of these innovations are already available, while others are starting to emerge. All of them have the potential to change how we live, work, and connect with our environment.

Take biochar, for instance. It might sound like something from a chemistry lab, but it's actually an ancient practice with modern implications. Biochar is made by burning plant matter in a way that locks carbon into the soil. Not only does this keep carbon out of the atmosphere, but it also makes the soil more fertile. Imagine a farmer in India or Kenya spreading biochar over their fields—crops grow healthier, the soil holds more nutrients, and all the while, carbon is being sequestered. It's a solution that tackles two problems at once: improving agriculture and fighting climate change.

Emerging technologies are pushing boundaries in exciting ways. A perfect example is direct air capture, which involves machines that take carbon dioxide straight from the air, helping to clean our atmosphere. While it's a new idea, with enough investment, it could scale up to remove billions of tons of CO_2 each year. Another promising development is green hydrogen. Unlike the hydrogen we produce now, which depends on fossil fuels, green hydrogen is created using renewable energy sources. It has the potential to power factories, cars, and homes, with water as its only byproduct. This kind of change could lead entire industries to become more sustainable.

Nature, too, is offering us some of the most innovative solutions. In coastal areas, mangrove forests are being restored to act as natural barriers against rising sea levels. These mangroves do more than protect our coastlines—they're great at capturing and storing carbon, too. Big restoration projects in places like Indonesia are helping to safeguard vulnerable areas while also pulling carbon from the air. It's a win-win, combining technology with the power of nature!

But for any of this to work, these technologies need to be adopted at scale. That's where governments, businesses, and communities come in. Policy incentives can encourage investment in these technologies, while public-private partnerships can bring the necessary resources to the table. Communities need to be engaged, too. Local farmers, businesses, and governments all have a role to play in making sure these solutions are integrated into everyday life. After all, technology is only as good as the systems that support it.

Economic Transformations

As new technologies take root and sustainability starts to redefine our norms, the way we produce and consume goods is also starting to change. The global supply chains that once prioritized speed and profit over everything else are now shifting toward something more sustainable. We're beginning to see a transformation in how products are made, used, and even disposed of—one that puts the planet front and center.

This shift is being driven by the circular economy approach. In a circular economy, nothing is wasted. The conventional "take-make-dispose" approach is replaced with products that are made to be recycled, repaired, and reused. Think about investing in a product that is meant to last for years rather than only a few seasons. And when it eventually reaches the end of its life, it doesn't end up in a landfill—it's reworked into something new. This model is already gaining momentum in places like the Netherlands, where cities are rethinking waste and focusing on how to keep resources in use for as long as possible. Japan is also at the forefront of material recycling, repurposing electronics and food scraps to create new, valuable items out of what would have otherwise been thrown away.

But why is this shift happening? The current system—the linear economy—has led us down a path where we take resources from the earth, make products, and then throw them away. It's a one-way street, and we're seeing the consequences everywhere. Landfills are filling up, oceans are choking on plastic, and the raw materials we rely on are becoming scarcer. This "use once and toss" mindset harms the environment and isn't good for the economy. Both businesses and governments are starting to see that there are better ways to move ahead.

Companies such as Patagonia and Unilever are setting an example of what this new way can look like. Patagonia is an outdoor gear company that, instead of encouraging the traditional cycle of buying a product and tossing it when it's broken, offers free repair services to customers so that they can extend the useful life of their goods. They've created a business model around sustainability, finding that doing good for the environment can also be good for business. Meanwhile, Unilever is one of the world's largest consumer goods companies and is rethinking how it does business. They've pledged to reduce plastic use, invest in recycling, and make their products have a smaller environmental footprint. These companies aren't following a trend; they're showing us what the future of business can be.

But this shift doesn't happen in a vacuum. It's being driven by changes in policies and regulations, alongside shifting consumer expectations. Governments are starting to implement policies that encourage businesses to adopt sustainable practices. In the European Union, for example, regulations are pushing companies to take responsibility for the entire lifecycle of their products, from production to disposal. At the same time, consumers are becoming more aware of the impact their choices have on the planet. People are starting to ask questions—where does this product come from? How was it made? What happens to it when I'm done with it?—and companies are being forced to answer.

To speed up this transition, we need more than just a few leading companies. We need economic policies that promote sustainability and discourage waste. We need rules that encourage industries to innovate and embrace circular practices. Most importantly, we need consumers to keep asking for better options. Every purchase we make is a vote for the kind of world we want to see.

Resilience and Adaptation

The truth is, we're shifting from just trying to stop climate change to learning how to live with it. While we aim to fix the damage, we also need to adapt to the changes that are already happening.

It's like trying to repair a ship while it's still sailing. We need to strengthen our communities, cities, and homes to prepare for what's coming.

Look at the Netherlands. They've faced flood threats for centuries and came up with a bold solution. They built the Delta Works, a massive system of barriers, dams, and locks. It wasn't only about keeping water out; it was about finding ways for cities to coexist with it. Take Rotterdam, a city that is mostly below sea level. Instead of building high walls, they created water plazas—open areas that fill with water during heavy rains to protect the rest of the city. It's smart and adaptable. This shows that we can do more than react; we can actually plan for the future.

On the other side of the world, in places like California and Australia, wildfires have become more than a seasonal threat. They're a serious concern that affects many lives. Every year, they destroy homes, wipe out forests, and displace entire communities. To adapt, people are learning how to build differently. Fire-resistant materials, like treated wood and metal roofs, are becoming part of the new normal. And in some areas, urban planners are leaving gaps—called firebreaks—to stop the flames from spreading too far. It's a simple idea, but it works. And in a world where temperatures are rising, simple solutions can make all the difference.

The idea is to work with nature instead of against it. The best solutions are often the ones nature has already created. One city that really gets this is Rotterdam. It's famous for its smart water management and serves as a great example of how cities can adapt to climate change. With green rooftops, rain-absorbing spaces, and a mindset that views water as a part of life rather than an enemy, Rotterdam shows us what adaptation truly looks like.

Rural areas are also finding their own ways to adapt. In places that face drought, communities are setting up systems to capture and store rainwater, making sure they have enough for dry times. In regions with extreme heat, planting trees for shade can mean the difference between a crop that thrives and one that fails. These simple solutions prove that adapting doesn't have to be complicated. This is the future we're heading toward—creating a world that is strong and resilient from the start instead of waiting to fix problems when they come up.

The Sociopolitical Landscape

Years ago, climate change wasn't a topic for political debate. It wasn't about countries making deals or developing nations influencing global plans. But today, climate change is one of the biggest challenges we face. It pushes every country to rethink how they live and how they lead.

Emerging economies like China, India, and Brazil are stepping into roles that no one predicted. The same countries that once stood on the sidelines are now central to the conversation. China, long criticized for its industrial emissions, is now the largest producer of solar panels and wind turbines. India, which still relies heavily on coal, is simultaneously ramping up its renewable energy ambitions with some of the largest solar farms on the planet. Brazil is balancing between the pressures of development and the international demand to protect the Amazon rainforest. The world is paying attention to these nations because their decisions affect not only their futures but also shape the future for everyone.

Countries like the U.S. and Europe used to be the main voices in climate discussions, but things are changing. Developed nations now need to do more than lead; they have to listen, too. Emerging economies have a big stake in the outcome, and the rules of climate talks are evolving. Beyond the numbers and policies about emissions and energy shifts, there's a more important conversation happening. It's about ethics. Are we leaving a planet that future generations can live on? Intergenerational justice urges us to think about more than just our immediate needs. We have to consider the rights of those who haven't been born yet. What right do we have to use up resources, pollute the air, and heat the planet when those who come after us will pay the price for our actions?

The ethical questions don't end here. Around the world, people are increasingly recognizing the rights of nature. New Zealand made headlines when the Whanganui River was given legal personhood. Now, the river isn't only a body of water; it has rights like a person. This bold idea is about more than protecting the environment; it's about acknowledging the connection we have with nature that we often forget.

Frameworks like the Earth Charter and the Universal Declaration of the Rights of Mother Earth push this idea even further. They remind us that to fight climate change, we need to change our view of nature from a resource to something that deserves our care. This is more than an environmental issue; it's about understanding what is right and wrong.

Building Momentum for a Sustainable Future

We understand what's at stake. The world we inherit depends on the choices we make today. Whether it's adopting new technologies, changing our economies, or considering our responsibilities to future generations, we get to choose the path forward.

Change starts in boardrooms, on farms, in neighborhoods, and in conversations about the future. Technological advances can improve our lives and how we use energy. Economic changes can help growth happen without harming our planet. Ethical governance can remind us that this is about fairness for both future generations and the Earth itself.

But the time to act is right now. Every step, from lowering carbon footprint to lobbying for better climate policies to backing green practices, advances us towards a more survivable – and sustainable– future. The world can still turn. The tide can turn. But it will take every one of us.

Mitigation Strategies and Solutions

CHAPTER

06

In this chapter, I present a literature review of what has been happening in climate mitigation strategies and solutions for the past decade. The ongoing impacts of climate change have underscored the urgent necessity for effective mitigation strategies and solutions as societies worldwide grapple with the ramifications of rising temperatures, extreme weather events, and ecological disruptions. Current academic discourse emphasizes the multifaceted nature of climate action, integrating approaches that span technological innovation, policy reform, and community engagement. While notable progress has been made in developing a diverse array of strategies, ranging from renewable energy adoption to carbon capture and storage, the effectiveness of these methodologies relies heavily on their context-specific application and the socio-economic frameworks in which they operate. Literature in this domain reveals a primary thematic focus on the interplay between mitigation efforts and broader socio-economic factors, highlighting how equity, accessibility, and politics will influence the success of implemented strategies. For instance, recent studies demonstrate that marginalized communities often face disproportionate burdens from climate impacts, necessitating solutions that are not only efficient but also equitable.

Furthermore, the role of governmental policies and international agreements, such as the Paris Agreement, has been widely examined, showcasing the importance of collaborative efforts at various levels—from local municipalities to global coalitions. Scholars have also explored the potential of innovative technologies, such as artificial intelligence and machine learning, in enhancing the efficiency of mitigation strategies. However, despite the extensive body of research, several critical gaps remain in understanding the dynamic interactions among various mitigation strategies and their long-term sustainability. Notably, there is an insufficient exploration of how individual behaviors and cultural factors shape the adoption and effectiveness of climate strategies, as well as a lack of longitudinal studies assessing the impacts of these strategies over time.

Moreover, the literature frequently overlooks the intersections of climate mitigation with biodiversity preservation, failing to adequately address how these dual objectives can be harmoniously integrated. As climate change continues to evolve, there lies an urgent need for interdisciplinary research that delves into these gaps, fostering a more holistic approach to environmental stewardship. This literature review aims to synthesize current knowledge on climate mitigation strategies and solutions, evaluating their strengths, weaknesses, and overall effectiveness while critically engaging with the identified gaps. The subsequent sections will systematically review existing studies, categorizing findings into technological advancements, policy frameworks, and community-centered approaches while also evaluating the significance of integration across disciplines. As we interrogate the current landscape of climate mitigation, it is imperative to not only highlight successful implementations and promising innovations but also to outline pathways for future research that could bridge existing knowledge gaps and promote actionable, scalable solutions. Thus, this review aspires to contribute to an informed and comprehensive understanding of the climate mitigation discourse, ultimately enhancing the collective ability to combat one of the most pressing issues of our time.

As climate change became increasingly recognized as a global issue in the late 20th century, the early 1990s marked a pivotal point in formulating climate mitigation strategies. The establishment of the United Nations Framework Convention on Climate Change (UNFCCC) in 1992 symbolized a collective acknowledgment of the necessity for international cooperation in addressing climate challenges (Digitemie WN et al., 2024). This was further solidified by the Kyoto Protocol in 1997, which aimed at reducing greenhouse gas emissions from developed countries, highlighting the role of legally binding targets as a cornerstone for climate mitigation efforts (Mpala TA et al., 2024).

The turn of the millennium saw a shift in focus towards enhancing technological advancements and renewable energy sources. Studies began to emphasize the importance of integrating renewable energy into national and international policies as a means to transition toward low-carbon economies (Murugesan R, 2024), (Afinjuomo OH et al., 2024). As countries recognized the socio-economic implications of climate change, frameworks that included green technologies gained traction, reflecting a growing intersection between economic development and environmental sustainability (Bano Q et al., 2024).

By the 2010s, the approach evolved to incorporate more local and community-driven strategies, such as Climate-

Smart Agriculture (CSA) and sustainable land management, which aimed to enhance food security and increase resilience to climate impacts (Rahmani H et al., 2024) (Oluwadamilare F et al., 2023). The Paris Agreement in 2015 marked another significant milestone, as it encouraged nations to adopt nationally determined contributions (NDCs) that encompass a broader range of climate mitigation strategies while promoting climate finance (A Abdallah, 2022), (Harper A. et al., 2018). This period also saw increasing recognition of the importance of carbon pricing mechanisms, which were framed as essential for incentivizing emissions reductions across various sectors (B E Law et al., 2018, p. 3663-3668).

In recent years, literature has begun to express the complexities surrounding equity and inclusivity in climate policies, pointing to the need for holistic approaches that concurrently address social justice and environmental sustainability (Hisano M et al., 2017, p. 439-456) (Untenecker J et al., 2016, p. 459-472). Consequently, ongoing research continues to explore innovative solutions aimed at fostering resilience and sustainable livelihoods, emphasizing the necessity for adaptive management strategies in the face of an ever-changing climate (Joanna I Lewis, 2021, p. 42-63), (Clora F et al., 2021). The evolving discourse around climate mitigation strategies underscores a shift from reactive measures to proactive, integrated approaches that seek to balance ecological integrity with human development (Batini N et al., 2020), (Ahmed M N Masoud et al., 2022, p. 3232-3232).

Climate mitigation strategies are increasingly focused on understanding the interplay between sustainable practices and greenhouse gas reduction. Central to this discourse is the concept of carbon pricing, which has emerged as a pivotal approach in numerous countries. By internalizing the cost of carbon emissions, carbon pricing mechanisms such as taxes and emissions trading systems incentivize businesses and individuals to reduce their carbon footprint (Digitemie WN et al., 2024). However, the effectiveness of these measures often hinges on accompanying policies that address social equity and competitiveness concerns (Mpala TA et al., 2024).

In parallel, the role of sustainable land management practices cannot be overstated. Empirical studies indicate that strategies like afforestation, reforestation, and the adoption of climate-smart agriculture not only sequester carbon but also enhance food security and biodiversity (Murugesan R, 2024), (Afinjuomo OH et al., 2024). For instance, integrating renewable energy into agricultural practices has shown significant promise in reducing emissions while simultaneously supporting rural economies (Bano Q et al., 2024).

Moreover, international cooperation is essential in the pursuit of comprehensive climate mitigation. Collaborative efforts through global agreements such as the Paris Accord have facilitated a unified approach towards emission reduction targets (Rahmani H et al., 2024). These agreements often encourage the sharing of technologies and best practices across borders, thereby reinforcing the necessity of collective action in the climate realm (Oluwadamilare F et al., 2023).

While exploring the nexus between economic growth and sustainability, understanding the impact of policies that promote renewable energy deployment can lead to significant reductions in carbon emissions (A Abdallah, 2022), (Harper A. et al., 2018). Overall, the integration of these diverse strategies highlights the multifaceted nature of climate mitigation, demonstrating that effective solutions require a combination of regulatory, technological, and societal shifts.

Different methodological approaches have significantly influenced the development and implementation of climate mitigation strategies and solutions, shaping how researchers assess effectiveness and feasibility. Quantitative methods have dominated the discourse, providing robust statistical analyses that enhance predictive modeling of emission reductions through various interventions. For instance, studies utilizing econometric modeling highlight the effectiveness of carbon pricing strategies in reducing greenhouse gas emissions, demonstrating correlations between price changes and emission reductions across different sectors (Digitemie WN et al., 2024). Conversely, qualitative methods emphasize the socio-political contexts behind these strategies. Research employing case studies often uncovers the barriers faced by marginalized communities in adopting renewable energy sources, revealing how socio-economic factors interplay with policy efficacy (Mpala TA et al., 2024).

Mixed-method approaches have gained traction, integrating quantitative and qualitative frameworks to capture the complexities of climate policies. For example, studies combining economic analysis with stakeholder interviews illuminate how perceptions of fairness could affect public acceptance of mitigation policies, such as carbon taxes (Murugesan R, 2024). Additionally, participatory research methods have emerged as vital in understanding local adaptation techniques in vulnerable regions, addressing both environmental and socio-economic dimensions of climate resilience (Afinjuomo OH, et al., 2024). By juxtaposing different methodologies, researchers can explore multifaceted responses to climate change while highlighting the importance of context-specific solutions.

Furthermore, utilizing Geographic Information Systems (GIS) has enabled researchers to spatially analyze land use changes linked to mitigation strategies. This approach assesses how afforestation efforts impact carbon sequestration potential across varied ecological zones (Bano Q et al., 2024). Each methodological lens contributes invaluable insights into the broader discourse on climate mitigation, underscoring the necessity for a diversified research agenda capable of addressing this complex global challenge effectively.

The pressing need for effective climate mitigation strategies has galvanized various theoretical perspectives that offer distinct frameworks for understanding and addressing this complex issue. For instance, ecological modernization theory posits that economic growth and environmental sustainability can be reconciled through technological advancement and regulatory reforms, supporting approaches like carbon pricing and renewable energy investments (Digitemie WN et al., 2024). This perspective highlights how innovative solutions can drive down emissions while promoting economic benefits, suggesting that a shift towards sustainable technologies can be both feasible and profitable.

Conversely, the political economy perspective critiques the assumption that market mechanisms alone can effectively address climate issues, emphasizing the influence of power relations and institutional frameworks in shaping environmental policy (Mpala TA et al., 2024). Scholars in this domain argue that without addressing underlying social inequalities and institutional barriers, climate mitigation efforts may disproportionately burden marginalized communities, thereby complicating sustainability goals (Murugesan R, 2024). This perspective is further supported by the argument that collective action is essential for effective climate governance, which the collective action theory underscores; it suggests that collaborative efforts across borders are necessary to combat global climate change successfully (Afinjuomo OH et al., 2024).

Moreover, sociological perspectives, such as the social construction of technology, suggest that the adoption of certain mitigation strategies is influenced by societal values and norms, thereby shaping public acceptance and implementation (Bano Q et al., 2024). This variation in theoretical approaches indicates that effective climate mitigation requires a multifaceted understanding that integrates technological, political, and social dimensions. By advancing a composite view that synthesizes these theories, researchers can better inform policy development and public engagement strategies aimed at achieving comprehensive climate action (Rahmani H et al., 2024).

The exploration of climate mitigation strategies and solutions has yielded significant insights that underscore their complexity and critical importance in addressing the multifaceted challenges posed by climate change. The findings reveal a consensus on the necessity of integrating technological innovations, policy frameworks, and community engagement, highlighting that effective mitigation cannot rely solely on any single approach. Renewable energy adoption, carbon pricing mechanisms, and sustainable land management practices have emerged as pivotal components in crafting effective responses to climate challenges, while the importance of equity and justice in these strategies has garnered increasing attention. Notably, the literature emphasizes the need for context-sensitive applications that consider local socio-economic conditions and the unique vulnerabilities faced by different communities, thereby reconnecting climate initiatives to grassroots realities.

The primary theme of this review affirms that climate mitigation is not merely a technical challenge but represents a complex interplay of social, economic, and political factors that require holistic engagement. Through the synthesis of various methodologies and theoretical perspectives, it becomes evident that an interdisciplinary approach is essential for advancing our understanding and efficacy in climate strategies. The nuanced comprehension of how different strategies intersect— ranging from international regulatory frameworks to localized community actions—articulates a growing recognition of the need for collaborative frameworks that enhance resilience while fostering socio-economic development.

Broader implications of these findings extend to the fields of environmental science, policy-making, and sustainable development. By emphasizing the interconnectedness of climate and socio-economic factors, these strategies have the potential not only to mitigate greenhouse gas emissions but also to drive innovation and economic growth in sustainable ways. The integration of diverse perspectives can facilitate the design of more inclusive policies that not only address climate change but also advance social equity, ultimately leading to more just and resilient societies.

However, the literature review also identifies notable limitations in current research, including a relative scarcity of longitudinal studies that evaluate the long-term impacts of various mitigation strategies. Furthermore, while there is rich discourse on technological solutions, there remains insufficient exploration of behavioral and cultural factors influencing individual and collective climate actions. Addressing the existing gaps calls for further inquiry into the interactions of climate mitigation strategies with local contexts and an examination of the socio-political barriers to their implementation. Future research could also prioritize interdisciplinary studies that examine the synergies between climate mitigation, biodiversity preservation, and sustainable development, thus paving the way for comprehensive approaches that recognize the intricacies of both ecological and human systems.

In conclusion, the convergence of research in climate mitigation strategies provides a robust foundation for further exploration and action. By fostering cooperation across sectors and integrating scientific insights into policy frameworks, stakeholders can better navigate the complexities of climate mitigation, ultimately enhancing the resilience of societies against the myriad challenges posed by climate change. The urgent need for innovative and equity-centered solutions remains clear, underscoring a collective responsibility to pursue comprehensive approaches that not only mitigate climate impacts but also promote sustainable and equitable development for all.

Eco-Anxiety and Climate Action

As the sun began its slow rise, casting golden light upon a world weighed down by the promise of its own undoing, a question settled heavily in the minds of those awake to its call. How does one live in a time where the Earth itself seems to tremble under the weight of human hands, where the very air carries whispers of a future undone by fire, flood, and famine? This is no longer a far-off reckoning confined to the imaginations of prophets or poets but a tangible, living fear—eco-anxiety, the unspoken shadow that grips the hearts of the young, the old, and all those who stand in its path. For a generation that has never known a time without the looming specter of climate change, this fear is not something distant or abstract. It is the pulse in their veins, the breath in their lungs, ever-present and relentless. Each headline, each warning of rising seas and burning forests is a reminder of a world unraveling at the seams. This anxiety is not merely felt—it is lived. A sensation as constant as the winds that carry the scent of drought across parched lands or the rains that fall heavier each year.

Even though it may be a new term, eco anxiety has been felt by many who have seen the slow destruction of the environment. It is a deeply human and unwavering response to the danger not only to the Earth but also to the delicate interconnectedness of all life. From young people seeing their future get smaller each season to Indigenous communities who rely on the land for survival, this feeling of anxiety impacts a wide range of people. However, some individuals experience the tremors most intensely, particularly the younger generation. Research has indicated that the current generation, entering a world already struggling with its own downfall, bears a heavier burden than many others. They received a planet on the brink of destruction, bringing with it a fear too heavy for any child to carry. The sensation of powerlessness can be overpowering for a lot of people, a profound and persistent fear that increases with each additional fire, flood, or storm. For some, it appears as anger - a justified rage toward the ancestors who watched as the world edged toward disaster.

However, not just young people experience this burden of anxiety. Indigenous people, who have ancestral lands rich in biodiversity, have an intimate understanding of this profound fear. To them, the danger is not only to the territory but also to their way of living and the fundamental character of their societies. As caretakers of the planet, they have been cautioning about the perils of disregarding the equilibrium of nature, and now they witness the rising waters and deforestation, seeing their worries come true. The weight of this eco-anxiety should not be handled silently. Therapy groups are popping up worldwide, providing a supportive environment for individuals grappling with overwhelming fears of environmental disaster. In this peaceful environment of communal areas, individuals gather not only to express their worries but to change them. The darkness of anxiety does not signify the end but rather the start of something new. In these groups, people are taught to transform their fear into activity and their anxiety into determination. They are not by themselves, and by acknowledging that together, they draw power.

Think about the narratives of individuals who have transformed their anxiety into a motivation for making a difference. Xiuhtezcatl Martinez, a youthful Indigenous campaigner, speaks with the passion of someone who has their future in jeopardy. Greta Thunberg, who transformed her individual protest into a worldwide campaign, turned her fear into powerful action by refusing to comply with the current situation. These numbers, along with many more, demonstrate that confronting eco-anxiety can lead to significant change.

Corporate Responsibility and Wellness

Let's discuss an exciting development taking place in businesses nowadays. Businesses are increasingly recognizing the importance of not only pursuing profits but also prioritizing the well-being of their employees and the environment. It is no longer just optional; it is now integral to the operation of successful companies. And the exciting thing is that they are finding ways to combine climate concerns with their employee wellness programs and corporate responsibility initiatives. It is a beneficial situation for both parties - employees' health and the health of the planet. Consider a company such as Patagonia. Their reputation for sustainability is widely recognized, however, what is particularly fascinating is how they integrate this same approach into their health initiatives. Providing gym memberships or free snacks is not the only thing that matters. They advocate for their staff to engage in outdoor activities such as hiking, biking, or participating in environmental initiatives. While their workers improve their physical fitness and manage their stress levels, they are also immersing themselves in nature, aligning perfectly with Patagonia's goal of environmental stewardship. It's similar to achieving two goals at once – but without causing harm to any birds!

Nike is following a similar path but putting its own unique spin on it. "Employees can now take time off to recharge with the new "well-being days" introduction. It is a day focused on both mental and physical well-being, but what makes it unique is how they connect it to their broader environmental objectives. Nike is making a strong effort to decrease its impact on the environment, and by assisting their employees in taking breaks and recharging, they are fostering a workforce that is better equipped to support these initiatives. The connection between healthy individuals and a healthy environment is undeniable.

It's not just about the company's internal operations. Numerous companies are actively engaging their employees in sustainability initiatives. Certain companies provide employees with paid time off to engage in volunteer work, allowing them to support environmental causes that are important to them. Picture taking a day off from work to participate in tree planting or beach clean-up activities! It is an excellent method to create an impact and foster a feeling of company pride. Individuals feel a stronger bond when they are aware that they are employed by an organization that aligns with their beliefs. Technology also has a significant impact. Businesses are currently utilizing various tools to monitor the effectiveness of their wellness and sustainability initiatives. They have the ability to observe the immediate effect these programs have on the health of employees and the environment. And can you believe it? When employees observe their company prioritizing this issue, they are more inclined to stay and contribute to resolving it. It starts a beneficial loop - individuals feel appreciated, they become healthier, and they strive for the company's success.

Certainly, leadership must take charge for any of this to occur. When leaders demonstrate concern for the environment and their employees, it establishes the company's overall attitude. They are the ones who are responsible for important choices, such as transitioning to sustainable energy or constructing environmentally conscious workspaces. When employees witness their leaders practicing what they preach, it motivates them to join in.

Another important factor to consider is that employees, particularly the younger workforce, are beginning to anticipate this from their employers. They desire to be employed by companies that prioritize the planet and take action to protect it. If businesses aim to draw in and retain top-notch employees, they must take initiative. This involves incorporating climate action into all aspects of their operations, including wellness initiatives and corporate responsibility efforts. Incorporating climate issues into employee wellness and CSR is not only a passing fad but also a glimpse into what lies ahead. Businesses that succeed in achieving this are creating stronger, more involved staff members and contributing to environmental conservation efforts simultaneously. And is that not the essence of success?

Climate Resilience

Today, cities are facing some of the most difficult climate challenges in history. However, the positive news is that they are discovering innovative methods to address these issues and enhance the quality of life for residents. Green rooftops, city parks, and intelligent water management systems are appearing in various locations, converting urban areas into livable, adaptable environments. These measures not only combat climate effects but also enhance the health and appeal of cities. Consider green roofs, for instance. These gardens are more than just green spaces on rooftops; they function as natural air conditioners, reducing the heat retained by cities and improving the energy efficiency of buildings. They absorb rainwater to decrease flooding and maintain air quality by filtering pollutants. Additionally, they provide us with small green spaces within the urban landscape, allowing individuals to unwind and experience a connection to the natural world.

Urban green spaces are experiencing a comparable impact. They are not just open spaces - they are areas to cool down in the heart of the city. Parks contribute to reducing temperatures by offering shade and releasing water vapor into the atmosphere. They also enhance air quality and provide individuals with a sanctuary from city life. Picture a park filled with children playing, adults taking a break from work, and the community gathering together to appreciate the outdoors. It is not just a park but also a representation of strength, both in terms of the environment and society. Proper water management is also an essential component of the larger picture. Urban areas are becoming more intelligent in managing rainwater by using sustainable drainage systems (SuDS) that imitate natural processes. These systems consist of features such as rain gardens and permeable pavements that decelerate water runoff and allow it to penetrate the soil, minimizing the chances of flooding. By improving water management, cities can enhance their ability to withstand extreme weather events and promote healthier living environments for all residents.

However, it's not solely focused on contemporary advancements in technology. There is much to be gained from the wisdom of indigenous peoples. Indigenous communities have been practicing sustainable land, water, and resource management for hundreds of years, predating the widespread concern of climate change. In the Eastern Himalayas, the Apatani people have mastered terrace farming and combining paddy fields with fish cultivation to both save water and promote biodiversity. These methods are not just environmentally friendly, they are also highly efficient. The

Irular tribe in the Western Ghats of India practices traditional pest control techniques and cultivates a variety of crops without using artificial chemicals. This ensures that the soil remains in good health and remains productive, which is crucial for long-lasting sustainability. By blending ancient wisdom with modern methods, a sustainable and resilient approach is created that can effectively tackle the challenges of our times.

Education is crucial in developing resilience as well. Schools are incorporating lessons on sustainability and climate resilience to encourage creative thinking about these topics among students. Picture classrooms where students are taught not just math and science, but also how to safeguard the Earth for future offspring. That is the type of schooling that gets all of us ready for a world that is constantly evolving.

Grassroots efforts also play a significant role in building resilience. Have you witnessed a community garden uniting neighbors to cultivate their own crops and tend to the earth? These little community initiatives foster a feeling of collective accountability and bring people together in significant ways. Resilience is not solely found at the highest level; it stems from lower levels and is fostered by caring communities. Financial instruments such as climate risk insurance and resilience bonds are also being utilized to assist communities in recovering from climate-related disasters. These are more than just financial instruments - they serve as crucial support, particularly for marginalized groups that are frequently most affected by weather-related incidents. It is crucial to guarantee that all individuals have access to these resources in order to create a just and strong future.

Resilience is not only about surviving but also about flourishing. It is an ongoing process that demands continual learning and adjustment. The solutions we are currently creating, whether involving green infrastructure, traditional knowledge, or financial backing, are building the groundwork for a future where cities and communities can resist the impacts of climate change. Building climate resilience is about preparing for the future while making life better today. Whether it's through green roofs, community gardens, or educating the next generation, every step we take brings us closer to a world where people and nature can thrive together.

Biodiversity and Ecosystem Services

One of nature's greatest tools in combating climate change is its own power. Forests, wetlands, and coastal ecosystems play a crucial role in absorbing carbon and offering habitats for numerous species beyond being simply beautiful landscapes. Restoring these ecosystems not only captures carbon but also revives areas in need, boosting biodiversity and fortifying the planet's resilience. Forests have long been recognized for their capability to remove carbon dioxide from the air. By planting new trees and replanting in deforested areas, forests can be transformed into carbon sinks once more. Their responsibilities extend further than just that. Forests provide habitats for numerous endangered species. Efforts aimed at enhancing forest management guarantee that forests can sustain biodiversity and store carbon effectively. An excellent illustration of this double advantage can be seen in the projects taking place on Amazon. Efforts to conserve and revive this rainforest are not just reducing climate change but also safeguarding homes for native communities and numerous species that depend on its wide area.

Wetlands and coastal ecosystems may not always be easily seen, but they are still crucial. Referred to as "blue carbon" ecosystems, mangroves, salt marshes, and seagrass meadows absorb carbon at a remarkable pace, sometimes even quicker than tropical forests. These environments function as sponges, absorbing carbon dioxide and protecting coastlines from storms and flooding. Mangroves serve as a crucial barrier for coastal communities, offering protection from increased sea levels and severe weather events. Look at the restoration projects being done in Southeast Asia for mangroves: as communities restore these regions, they are safeguarding their residences and rejuvenating marine life. Restored mangroves help boost fish populations, benefiting both the local economy and the ocean's well-being.

However, the work is not limited to just restoring ecosystems—protecting and linking them is just as important. This is the purpose of ecological corridors. Corridors connect separated habitats, enabling species to move, locate food, and reproduce. In the absence of these links, animals may be confined to isolated areas, resulting in inbreeding and reduced genetic variation. The Yellowstone to Yukon Conservation Initiative in North America is among the biggest projects aiming to establish these crucial connections. The initiative links 3,200 kilometers of protected areas to allow grizzly bears, wolves, and caribou to move freely across borders, helping to maintain their healthy and thriving populations.

Indigenous communities have held responsibility for the land for many years, and their expertise is incredibly helpful in current initiatives to improve ecosystems. Their sustainable efforts go beyond simply surviving - they are focused on coexisting harmoniously with the environment. In the Eastern Himalayas, the Apatani people are skilled in the practice of terrace farming. This approach helps save water and also enables the cultivation of crops and fish in a manner that enhances biodiversity. In the Western Ghats, the Irular tribe employs organic pest control practices to maintain soil fertility, enabling them to cultivate a range of crops sustainably. These methods provide answers that contemporary science now acknowledges as crucial for preserving ecological equilibrium.

Reviving and safeguarding ecosystems not only aids the environment but also provides a means to protect humanity. With climate change putting our systems at risk, rebuilding forests, wetlands, and coastal areas offers us a chance to fight back. These natural habitats assist in controlling the climate, shielding communities from severe weather, and supporting the thriving of wildlife. By creating ecological corridors and drawing from the knowledge of indigenous peoples, we can ensure that ecosystems are both restored and conserved for future generations.

What is thrilling is that these endeavors are starting to gain momentum. Recognition of the effectiveness of nature-based solutions is increasing, from reforestation projects in the Amazon to wetland restoration along coastlines. These efforts emphasize the significance of investing in the environment, not just to address climate change but also to protect biodiversity. They serve as a reminder that nature has the ability to bounce back and protect us if we allow it to heal itself.

Community-Led Solutions

Around the globe, local populations are seizing control—quite literally—of their own power. These communities are producing their own energy by utilizing renewable sources such as solar, wind, and biomass, thereby decreasing their dependence on conventional fossil fuels. However, in addition to simply ensuring that the lights stay on, these initiatives are giving individuals a feeling of authority in managing their energy consumption and money.

For example, consider Hoboken, New Jersey. This city initiated a community solar program with the goal of offering clean, affordable energy to residents with lower and middle incomes. The program assists households in reducing their electricity costs and puts the money saved toward sustainability projects chosen by the community. It is a system of mutual rewards - where each person contributes, and each person reaps the benefits. In rural communities, wind power is causing a comparable wave of transformation. Communities that were previously dependent on declining income from agriculture are now profiting from wind farms, which produce not just electricity but also employment opportunities and additional sources of tax income. This active participation from the local community guarantees that the economic benefits are experienced by everyone in the area. These initiatives are not only providing energy independence but also revitalizing local economies, demonstrating that renewable energy is not just an environmental option but also a feasible financial one.

In terms of agriculture, communities are embracing sustainable methods that cooperate with nature instead of going against it. For instance, Permaculture is being adopted for its capacity to develop self-sufficient ecosystems. In France, the Bec-Hellouin farm showcases permaculture principles in practice. The farm flourishes through the integration of various crops and livestock, which enhances biodiversity and enhances soil quality. Even more impressive is the fact that this small farm has turned into an international center for education, proving that sustainable farming can be successful in any location.

Organic farming is also an essential part of the overall picture. Organic farmers protect the land and water, preserving soil fertility for future generations by steering clear of synthetic chemicals. However, in addition to the advantages for the environment, organic farming also fosters a sense of community among people. Farmers' markets and community-supported agriculture programs (CSAs) are thriving, providing opportunities for individuals to come together around fresh, nutritious food. These programs facilitate the growth of local economies and encourage a stronger sense of community. It's not just about purchasing food - it's also about establishing connections with the farmers who produce it.

Regenerative agriculture goes beyond sustainability by actively focusing on enhancing soil health. The Balbo Group in Brazil is leading the way in sugarcane farming by implementing methods that involve returning organic material to the soil and utilizing natural pest control strategies. This not only enhances the land's capacity for crop production but also aids in trapping carbon, making it a valuable weapon in combating climate change. The advantages are evident: improved soils lead to higher crop yields and decreased environmental impact.

Both community-driven energy and agriculture projects are united in their approach to developing local responses to worldwide challenges. Communities are increasing their resilience to environmental challenges by generating their own power and implementing sustainable farming practices that preserve the earth. It's not only about staying alive but also about forming more cohesive communities that are actively shaping their own destinies. These grassroots movements demonstrate that significant change can begin without the involvement of big organizations or governments. Frequently, it starts within the heart of nearby neighborhoods, where individuals are uniting to create an impact. By implementing renewable energy projects and sustainable farming practices, communities demonstrate that their actions can motivate worldwide change. This is the future of sustainability, driven by and for people.

Public Engagement and Behavior Change

Narratives have always held significant power. They influence our perception of the world, our emotions, and our behavior. When discussing climate change, the use of media, art, and community events is crucial in changing public attitudes and promoting significant behavior changes. While climate change may seem far off or unclear, a compelling story can make it feel personal and urgent, capturing our attention.

Consider, for example, the influence of movies such as An Inconvenient Truth. It went beyond just providing information, actually conveying the sense of immediacy surrounding the climate emergency. In the same way, Rachel Carson's Silent Spring motivated environmental movements by demonstrating the actual effects of human behavior. In more recent times, initiatives such as the #ImPactWithYouth videocast series have brought this to a more intimate level, showcasing narratives of individuals impacted firsthand by climate change. These stories help people connect to the problem, creating compassion and encouraging action by showing how similar individuals are impacted by the crisis.

Art also has a significant impact on this movement. Visual storytelling has the ability to elicit emotions that data and graphs cannot do. Artists such as Olafur Eliasson design installations that plunge viewers into the impacts of climate change, like increasing sea levels and melting glaciers. When standing in front of one of his artworks, observers not only witness the impact but also experience it. Art has the power to simplify complicated problems into something tangible, prompting individuals to contemplate their own behaviors and ways in which they can contribute to resolving them.

Storytelling plays a role in bringing people together through community events. Events such as Woods & Wilds blend art, storytelling, and music to connect individuals with nature and climate concerns in a laid-back, pleasant environment. These occurrences facilitate dialog and contemplation, paving the way for change in a manner that resonates with all individuals.

Education is also a crucial instrument for fostering lasting involvement. Schools and universities are enhancing their education by incorporating climate action into their curricula, preparing students with the necessary knowledge and abilities to have an impact. Anant National University in India's Climate Champion Program enables students to take on leadership roles in climate action in their communities. They are not only acquiring knowledge - they are putting that knowledge into practice in real-life scenarios. This interactive method transforms students into champions, prepared to address the environmental obstacles ahead of us.

In Kerala, the Student Police Cadet project goes beyond climate education by integrating environmental awareness into wider civic education. Students are not only studying the environment; they are also engaged in activities such as planting trees, managing waste, and conserving resources. It's a pragmatic approach to integrating climate action into daily life, embedding values that students will carry with them beyond their school years.

These school programs are backed by projects initiated by students, which are causing positive changes in educational institutions. Climate clubs and environmental organizations are emerging everywhere, providing students with opportunities to collaborate on initiatives such as decreasing energy consumption in their schools or promoting eco-friendly practices on campus. These initiatives extend beyond the school premises, reaching out to the broader community and impacting families, local businesses, and policymakers. When students assume responsibility for climate action, it leads to a ripple effect that promotes more widespread change.

Together, storytelling and education wield significant power. By making climate change personal through narratives and empowering youth with resources and chances to make a difference, we're not only increasing awareness but also igniting significant, enduring transformation. This method links emotions and intellect to tackle current climate change issues and create a basis for a positive, long-lasting future.

The Role of Technology and Innovation

The Empathy Imperative

As the climate crisis intensifies, it has become crucial for empathy to drive significant action. Technology plays a vital role in boosting worldwide empathy by providing immersive experiences that help people grasp the effects of climate change. An impactful demonstration is virtual reality (VR), immersing users in environments confronting climate dangers, making the emergency feel immediate and pressing. One example is Murdoch University's VR simulation, where individuals can feel what it's like to be a tree in a forest facing deforestation. Users experience the destructive impacts of a forest fire in this engaging setting. The encounter goes beyond just a video or statistic; it involves facing the impacts of environmental degradation firsthand. Research has indicated that immersive experiences of this nature have a lasting impact on individuals, motivating them to embrace more sustainable practices.

Another instance showcasing VR's ability to generate empathy is demonstrated by the collaboration between the United Nations Environment Program (UNEP) and Sony PlayStation. Collaboratively, they developed a virtual reality (VR) simulation that allows users to see their personal carbon footprint. In this simulation, the abstract idea of carbon emissions materializes as a huge orange gas ball that expands as the user interacts, never leaving their side. This straightforward, visual depiction of emissions encourages users to reconsider their daily behaviors and understand their contribution to the larger issue of global warming.

However, virtual reality alone does not cultivate empathy. Social media platforms are now effective tools for promoting action on climate change by amplifying voices and forming communities. Twitter, Instagram, and TikTok are now essential platforms for people and businesses to share their narratives and knowledge, connecting with audiences worldwide. An excellent illustration is Greta Thunberg's Fridays for Future movement. A single school strike has evolved into a global movement, mostly fueled by social media. These platforms enable instant conversation, transcending boundaries and promoting a feeling of collective immediacy. Millions of individuals have been motivated by Greta's narrative to become part of the movement, demonstrating that social media can play a crucial role in creating change.

Art also plays an important part in promoting empathy. The ability of visual art to elicit emotions is unique and cannot be replicated by statistics or data. Artists such as Olafur Eliasson have designed installations that render climate change tangible and urgent. His installations, like the melting ice blocks displayed outside the Tate Modern, fully engage spectators in the stark truth of increasing temperatures and disappearing glaciers. Art connects with audiences emotionally, prompting them to contemplate their involvement in both the issue and its resolution. Besides technology and art, education plays a vital role in empowering the future generation to make a difference. Educational institutions are now integrating climate change into their course content more often, with a focus on hands-on learning experiences that extend outside of traditional classroom settings. The Climate Champion Program at Anant National University in India serves as an ideal illustration. By participating in community projects that specifically target local environmental issues, this program motivates students to actively promote sustainability. By engaging students in practical, real-life projects, these programs make sure that young individuals are not only gaining knowledge about climate change but also taking the lead in addressing it.

Another remarkable effort is the Student Police Cadet program in Kerala, India, which combines environmental education with larger civic duties. By engaging in tree planting and waste management, students develop the skills to be responsible caretakers of the environment. These encounters instill a deep sense of obligation and equip young individuals with the necessary skills to guide their communities toward sustainability. Together, these technologies and educational programs showcase the profound impact of empathy. Through virtual reality experiences, utilizing social media, involving individuals with art, and providing students with practical experience, we are creating a worldwide community that recognizes and senses the importance of taking action on climate change.

In a society where facts and figures are frequently insufficient to inspire action, empathy is increasingly seen as the link between awareness and impactful transformation. As these technologies and programs driven by empathy develop further, they have the potential to not just inform but also motivate a joint effort to address the biggest challenge of our era: protecting our planet for the coming generations.

The Green Metamorphosis

Today, cities are presented with a distinct challenge: how to expand and progress without harming the environment. This is when the idea of eco-cities becomes relevant, offering a plan for changing cities that emphasizes sustainability in all aspects. The transition to eco-cities involves more than just cutting emissions or increasing tree planting—it requires a complete rethinking of how cities function, starting from the basics. By integrating eco-friendly technologies with intelligent urban planning, cities are able to transition in a way that prioritizes both the well-being of individuals and the environment like never before.

Copenhagen, Denmark, offers a compelling example of what this transformation may look like in actuality. Known for its ambitious aim to achieve carbon neutrality by 2025, the city stands as a model of innovative urban planning. Nevertheless, Copenhagen's advancement is not solely dependent on a top-down approach. The city's success is attributed to its integration of environmentally conscious practices into everyday activities. Examine the biking infrastructure in Copenhagen, which has been heavily invested in creating a network of bike highways, encouraging the use of bicycles as a primary mode of transportation. The city's well-maintained and well-illuminated streets are connected to provide residents with a convenient option for cycling instead of driving. This not only reduces the city's emissions but also improves the general health and wellness of its inhabitants. It demonstrates how focusing on one specific strategy, such as promoting biking, can have extensive impacts.

Another notable aspect of this transformation is the Nordhavn district in Copenhagen. Nordhavn, previously an industrial harbor, has been transformed into a "smart city energy lab," demonstrating the integration of renewable energy sources like wind and solar into urban life. The heating and electricity systems in the district are coordinated to function together, maximizing energy efficiency. Furthermore, structures in Nordhavn are built with the aim of enhancing energy efficiency, incorporating features such as rooftop solar panels and advanced insulation methods. This area isn't only focused on reducing environmental harm—it's about demonstrating how urban areas can prosper in a low-carbon tomorrow.

Singapore has tackled its urban transformation by transforming its dense, vertical cityscape into a lush green haven. Singapore has led the way in vertical gardens and green buildings due to limited space, showing that even densely populated cities can connect with nature. The Supertree Grove at Gardens by the Bay is possibly the most recognizable representation of this idea. Dominating the city skyline, these towering supertrees are not just a feast for the eyes, but they also have a practical function. They have solar panels, rainwater collection systems, and cooling ducts, which allow them to function as ecosystems. Singapore's dedication to sustainability extends beyond these prominent projects; the city has also put in place large-scale tree-planting programs, waste-to-energy facilities, and a sophisticated water recycling system.

At the same time, Freiburg in Germany shows that the shift to green practices involves both community and technology. In the Vauban neighborhood, citizens play a crucial role in the city's sustainability objectives. Residences in this area generate more energy than they use, making them energy-positive. What sets Vauban apart is its community-driven design: residents actively participate in planning decisions, and car usage is discouraged to encourage walking, cycling, and using public transportation. The strong ownership felt in Freiburg has led to successful sustainability efforts, emphasizing the significance of community involvement in shifting toward green practices. In this place, urban change is not forced upon the people - they play an active role in shaping it.

Nevertheless, transitioning from conventional cities to eco-cities is not free of difficulties. Upgrading current infrastructure to adhere to contemporary sustainability norms may involve high expenses and intricate procedures. Many cities struggle with how to include smart grids - sophisticated systems that enable efficient energy use and seamless integration of renewable energy sources. These grids allow cities to oversee electricity usage in real time, preventing energy wastage and enabling the storage or sharing of excess renewable energy within the city. This type of innovation is crucial in enabling cities to meet future energy needs while decreasing their dependence on fossil fuels.

Another crucial element of the puzzle is transportation. Electric buses, bike-sharing programs, and pedestrian-friendly city designs are sustainable transport systems that are aiding in the reduction of greenhouse gas emissions and enhancing mobility. Electric buses are now common in cities such as Amsterdam and Oslo, reducing pollution in city centers and providing residents with a cleaner, quieter mode of transportation. Simultaneously, bike-sharing programs are growing in popularity in cities worldwide, enabling residents to easily rent bikes for brief journeys. These efforts demonstrate the potential for transportation to adapt in order to promote sustainability while maintaining convenience.

The transformation of cities into green spaces involves more than just technology or infrastructure; it also entails altering people's lifestyles and relationships with their surroundings. Cities such as Copenhagen, Singapore, and Freiburg show that implementing intelligent urban design and sustainable technologies can significantly decrease environmental harm and enhance quality of life. The outcome is a novel form of city – one that is not only resistant to climate change but also built to promote healthier, happier communities. This change signifies a basic reconsideration of the concept of constructing and residing in urban areas. As more cities adopt this green change, they lead the way for a future where urban areas are both sustainable and flourishing with nature.

Innovative Water Management Technologies

Innovative water management technologies are becoming crucial due to rising global water demand and decreasing reliability of traditional sources caused by climate change. These progressions involve more than just discovering new water sources; they involve utilizing existing water more effectively. These technologies are creating opportunities for more sustainable practices in both urban areas and agricultural fields, effectively preserving and controlling water with accuracy and effectiveness.

An innovative technology is the extraction of water from the atmosphere. This technique converts water vapor in the air into purified drinking water by removing moisture. It is a solution that seems futuristic, but it is already being used in areas with water shortages. Businesses such as HurRain NanoTech in China are leading the way in this technology, utilizing advanced substances to enhance adsorption and enable water collection in regions with low humidity. In arid regions with unreliable or overused traditional water sources, atmospheric water capture powered by solar energy provides a sustainable and renewable solution to secure water for communities. As technology advances, the potential to address water shortages on a larger scale increases, offering hope for improved water security.

An innovative technology is the extraction of water from the atmosphere. This technique converts water vapor in the air into purified drinking water by removing moisture. It is a solution that seems futuristic, but it is already being used in areas with water shortages. Businesses such as HurRain NanoTech in China are leading the way in this technology, utilizing advanced substances to enhance adsorption and enable water collection in regions with low humidity. In arid regions with unreliable or overused traditional water sources, atmospheric water capture powered by solar energy provides a sustainable and renewable solution to secure water for communities. As technology advances, the potential to address water shortages on a larger scale increases, offering hope for improved water security.

The agricultural industry is being changed by smart irrigation, which merges data-driven observations with live monitoring. Conventional methods of irrigation may result in the wastage of water, whereas smart irrigation ensures that every drop is utilized efficiently. Devices are inserted into the ground to monitor moisture levels, while algorithms analyze weather predictions and crop needs to adjust irrigation timings accurately. This indicates that farmers are able to provide their crops with the precise amount of water needed without excess or deficit. Smart irrigation is advantageous because it conserves water, increases crop production, and reduces expenses. As more farmers start using these systems, we are noticing substantial enhancements in both water saving and farming output.

Digital water management technologies are also crucial, especially in urban areas with intricate infrastructures that need close monitoring. Through the utilization of AI and IoT sensors, municipalities can continuously monitor their water infrastructures. These tools enable rapid leak detection, accurate water usage monitoring, and efficient resource distribution to ensure water is delivered to where it is most needed without any wastage. Real-time management is essential for cities with aging infrastructure or high population densities, as minor inefficiencies can result in substantial water losses.

The way we view water management is changing with the integration of these technologies. We are no longer exclusively concerned with locating new water sources; now, we are improving our usage of existing water. By collecting moisture from the atmosphere, reusing wastewater, and utilizing intelligent irrigation techniques, we are developing systems that can better withstand the difficulties of water scarcity. These advancements bring optimism for a future where smart planning and innovative solutions help tackle water scarcity issues.

As these technologies become easier to access, there is a greater opportunity to expand their global impact. Urban areas, agricultural lands, and industrial sites all have the potential to gain advantages by implementing more advanced techniques for managing water. The important thing now is to make sure these solutions are implemented on a large scale in order to have an impact. By using water responsibly and consistently innovating, we can ensure water availability for future generations and reduce negative effects on the environment.

Energy Revolution: Advances in Clean Energy Technologies

The world's energy situation is changing, with a focus on clean energy technologies becoming prominent in the move toward a more sustainable tomorrow. As the world looks for options other than fossil fuels, advancements in solar, wind, and newer areas such as tidal and wave energy are propelling this change. These advancements are not just transforming power generation methods but also bringing us closer to a future fueled by renewable energy sources.

Solar power continues to be a fundamental aspect of this change due to continuous enhancements in its effectiveness and adaptability. Leading the way are perovskite solar cells, which have the incredible ability to surpass conventional silicon-based panels. Perovskite cells demonstrate high efficacy in capturing sunlight, recording conversion rates exceeding 33% in laboratory experiments. When paired with silicon in dual configurations, they are able to capture a wider range of the light spectrum, expanding the capabilities of solar cells. These developments enable higher power generation from the same sunlight quantity, increasing the efficacy of solar energy. Thin-film solar cells are also transforming the market. Crafted with substances such as copper indium gallium selenide (CIGS), these cells are light, bendable, and suitable for installation on various surfaces, not limited to just roofs—they can be incorporated into the exteriors of buildings, vehicles, and even handheld electronic devices. This flexibility creates new opportunities for solar energy utilization by converting previously unutilized surfaces into energy-producing resources.

Wind power is also reaping the rewards of fast technological progress. Turbines are becoming bigger and more effective, gathering a greater amount of energy than in the past. Contemporary wind turbines are created with longer blades and taller hub heights, enabling them to harness more powerful and steady winds present at elevated levels. This results in increased energy production, leading to improved efficiency and cost-effectiveness of wind farms. Offshore wind power, specifically, is experiencing substantial expansion. Floating wind turbines are harnessing the potential of deeper ocean waters, where fixed foundations are not viable but wind conditions are optimal. Offshore farms are utilizing these strong winds to generate a constant supply of renewable energy. Progress in materials and aerodynamics is enhancing the performance of turbines. Contemporary blade designs are created to endure tough conditions and increase energy capture, with new materials advancements leading to lower maintenance and operational expenses.

The oceans contain an additional undiscovered supply of sustainable energy: tidal and wave power. Tidal power, as opposed to wind or solar power, is highly reliable. The ebb and flow of tides adhere to a consistent, dependable schedule, ensuring that tidal energy remains a reliable source of power. FastBlade and CoTide are at the forefront, creating turbines that effectively utilize tidal currents. These systems are created to function in tough marine conditions, guaranteeing longevity and optimizing energy extraction. Wave power is also progressing. Devices such as the WaveRoller created by AW-Energy change the movement of sea waves into electrical power. Wave energy is constant, provided by waves day and night, making it a dependable renewable energy source. While tidal and wave energy are not as developed as solar and wind power, they show great promise for coastal areas seeking to vary their energy sources.

Moreover, clean energy is becoming increasingly affordable alongside these technological progressions. Over the last ten years, the price of solar and wind energy has significantly decreased, allowing businesses, governments, and individuals greater access to renewable energy sources. Solar power has become one of the most inexpensive electricity sources in various regions worldwide. The decrease in expenses, along with advancements in storage devices such as lithium-ion batteries, is simplifying the process of storing and delivering renewable energy. This guarantees a steady supply even during periods of minimal sunlight or wind.

The consequences of this energy revolution go beyond just the environmental advantages. The increase in renewable energy is leading to the creation of millions of new jobs worldwide. The renewable energy industry is thriving, with engineers creating new wind turbines and technicians servicing solar farms. Often, these positions provide enduring, secure jobs in areas that typically depend on sectors such as coal and oil, supporting a fair transition for workers. Additionally, nations that invest in renewable energy are not just decreasing their carbon footprints but also achieving energy independence, lessening dependence on imported fossil fuels, and enhancing national security.

This energy transformation involves more than just technology; it involves reconsidering how we fuel the planet. Solar, wind, and ocean energy together are changing the worldwide energy scene, bringing us nearer to a future where sustainable, clean energy drives both our cities and homes. The ongoing advancements in renewable energy technologies are transforming a previously far-off idea into a tangible outcome. With the advancement and expansion of these technologies, there is a possibility of revolutionizing the way we live, work, and provide energy for our planet.

Biotechnology and Climate Solutions

Biotechnology is becoming increasingly important in the battle against climate change, providing creative solutions such as modified crops and engineered microorganisms for capturing carbon. These developments are changing our approach to sustainability, expanding the limits of what can be achieved in agriculture and environmental management.

Creating genetically modified crops to tackle climate change challenges is a significant progress in the field of biotechnology. Traditional farming often relies on pesticides and chemical fertilizers, leading to environmental damage and increased greenhouse gas emissions. However, genetically engineered crops have been created to resist pests and diseases, reducing the need for chemical products. An example of this is crops that resist insects, which have helped reduce overall pesticide use worldwide, leading to lower emissions from agriculture. Using fewer pesticides leads to less introduction of toxic substances into the soil, air, and water, as well as a decrease in the energy-intensive processes involved in their production and application.

In addition to pest resistance, genetically modified crops are being developed to flourish in challenging environments, like areas susceptible to drought. These crops that can withstand drought conditions are specifically developed to thrive with much lower water usage, aiding in the preservation of this crucial resource in regions facing scarcity. Through improving the efficiency of water usage in agriculture, biotechnology is actively tackling a key challenge presented by a shifting climate: guaranteeing food security while lowering the environmental impact of farming.

However, the effects of GM crops are much more important than many people realize. Herbicide-tolerant varieties are encouraging the uptake of no-till farming. The combination of being able to keep weeds at bay with the concomitant immense benefits of no-till farming – maintenance of soil structure and its avoidance of erosion, along with enhanced soil's ability to store carbon – means that GM crops foster not just yield in crops but also contribute to carbon sequestration, locking away carbon where it's needed, in the soil, rather than permitting it to escape into the atmosphere. This double benefit – more crops, less emissions – reveals biotech's promise to reorient agriculture toward a more sustainable future.

Innovation in biotechnology is also moving beyond crops: microorganisms – especially bacteria and fungi – are engineered to act as drivers in carbon capture and storage. Nature does this already: these microbial allies capture and store atmospheric carbon in soils more efficiently than many proposed strategies: today, synthetic biology researchers are building bacteria for carbon capture and conversion to useful products such as biofuels and bioplastics, with the triple benefit of removing CO_2 from the atmosphere, turning it into useful products and substituting fossil fuel-based products that rely on non-renewable resources.

Marine microorganisms can also be key contributors to the global carbon cycles. Photosynthetic microbes in the ocean – such as algae and cyanobacteria – can suck up large quantities of carbon dioxide as part of their natural life cycles. They can also add to carbon sequestration by being harvested and turned into biofuels. This is where some biotechnological approaches come into play. For instance, one could modify the cells of naturally photosynthetic microbes to make them more efficient in the capture of atmospheric carbon dioxide. If the scaling-up of these natural processes was done properly, it could become an important part of global carbon sequestration strategies, complementing land-based approaches.

These examples of biotechnological innovations show how thoroughly science and technology can overlap in practical responses to climate change; from water-saving solutions in agriculture to novel carbon capture technologies, biotechnologies are forging routes to resilience and sustainability. Now, these and other technologies need to be integral to strategies and policy responses to deal with the inevitable impacts of climate change.

With advances in biotech research and development, we are on track to actualize a future with stable, climate-resilient food production systems powered by novel biotech crops and microbes that can capture carbon.

Envisioning the Future CHAPTER 08

The Psychological Climate

When it comes to adopting sustainable practices, a great deal of the challenge is psychological: many of the barriers to forming more sustainable habits can't be overcome by better technology or policies alone but rather by how people feel about change. One of the largest mental barriers is status quo bias – a tendency to prefer things as they are – that can lead people to feel risk averse, even when they know that change would lead to a better outcome. Even when they know that a more sustainable choice is beneficial, many people will stick with the same routines they have always used.

This resistance is often rooted in another strong psychological factor: loss aversion. The notion that people weigh prospective losses more heavily than prospective gains tends to manifest in the face of potential sustainable behaviors. If reducing one's ecological footprint means giving up certain inconveniences or habits that might feel comfortable and safe, people will, more than not, focus on what they might lose rather than on the long-term benefits of a sustainable lifestyle. Even though the benefits of sustainable actions – for instance, less waste, less energy use or less emissions – might be obvious, the immediate perceived discomfort of change feels too big to bear. This type of thinking can deter people from embracing the solutions needed to address climate change.

On top of that, there's a widespread fear of the future. Climate change, while urgent, often feels distant to many people. The effects might be catastrophic, but they're often framed as problems for future generations or faraway places. This "it won't happen to me" mentality encourages short-term thinking, where the priority becomes dealing with today's challenges instead of making changes that protect tomorrow. This makes it easier to justify inaction, as the risks seem too remote to prompt immediate behavior shifts.

What can we do to conquer these challenges? Education is one of the most potent tools available. It goes beyond just sharing information; it entails empowering people to evaluate the world and their influence on it. By incorporating sustainability into their curriculum, schools and universities are teaching students while also providing them with the tools to challenge conventional methods and generate creative answers. People can make choices that are better for the environment by recognizing and acknowledging the cognitive biases that hinder them. This form of education lays the foundation for a shift in mindset that reaches beyond the classroom and into everyday life.

Community involvement is equally important. People tend to change their actions when they see others doing the same. When a community collaborates to back a sustainability effort like community gardens, recycling programs, or switching to renewable energy, it initiates a domino effect. Local authorities, instructional workshops, and events may show real examples of success, inspiring people to replicate that achievement. This kind of engagement encourages a sense of shared responsibility, showing that individual actions can impact a broader scale. When individuals join together, the shift toward sustainability is seen as a collective endeavor rather than a personal duty.

Another effective method for influencing behavior is employing "nudges." These are slight alterations in the surroundings that motivate individuals to improve their decisions with minimal additional effort needed. For instance, positioning recycling bins strategically or providing renewable energy as the automatic choice on energy plans can guide individuals toward more eco-friendly actions without seeming like a significant compromise. Simple, readily implementable measures can frequently produce a substantial influence when implemented broadly. Nudges are effective because they simplify decision-making, guiding individuals toward positive choices without causing stress.

However, it's not only about modifying individual actions. We must also tackle the cultural beliefs that hinder advancement. Sustainable choices are often perceived as "inconvenient" or "costly," contributing to the psychological obstacles we have addressed. Transforming this story necessitates a mix of learning, community guidance, and strong policy backing. When individuals realize that sustainability is not only achievable but also viable and advantageous in their everyday activities, it becomes simpler to address the psychological obstacles hindering change.

At the end of the day, surpassing these mental obstacles is not about coercing individuals to change but about establishing environments that facilitate making sustainable decisions that are more achievable and appealing. Education, involvement in the community, and strategic nudging collaborate to foster a sustainable culture. By confronting the fears, prejudices, and misunderstandings that hinder individuals, we can pave the way for a future where acting in an environmentally conscious manner is common practice rather than unusual.

Blueprints for a Green Tomorrow: Case Studies in Sustainability

Across the globe, there is a silent change happening in various communities, cities, and nations that have adopted sustainability as a requirement rather than a choice. These locations are taking the forefront in demonstrating how policies, technology, and cultural changes can combine to develop flourishing communities that benefit both individuals and Earth.

The path of sustainability has been shaped by important historical events and ideas, showing a changing view of how humans interact with the environment. It started gaining traction during the environmental crises of the 1960s and 1970s, building on the efforts of grassroots movements and global talks, especially the 1972 Stockholm Conference and the 1987 Brundtland Report, which introduced the phrase sustainable development. This report highlighted the need to balance taking care of the environment with social and economic growth, showing an early recognition of how ecological and human systems are connected. At the same time, efforts like the UN Decade of Education for Sustainable Development showcased the need for thorough educational methods, as seen in programs such as the Systems Thinking and Practice (STiP) from the Open University (Blackmore et al., 2015). These key moments demonstrate an understanding that sustainability includes not just environmental health but also social fairness and economic strength, creating a foundation for modern sustainable practices (Jesus et al., 2024).

A. Evolution of Sustainability Concepts Over Time

Over many years, different aspects of sustainability have come up as society faces issues like environmental harm, social unfairness, and economic problems. At first, sustainability was mainly seen from an ecological point of view, focusing on conservation and biodiversity. As understanding grew, the idea changed to include social fairness and economic stability, leading to the use of frameworks such as the United Nations Sustainable Development Goals, which show how these areas depend on each other. This change is shown in technology improvements, like the use of low and zero-carbon technologies, which help consumers engage in sustainable practices ((Caird et al., 2007)). Also, the growth of Building Information Modeling (BIM) shows how engineering and architecture are now focusing on sustainability through processes that include the entire lifecycle of buildings ((Aish et al., 2019)). This diverse development shows a key shift towards comprehensive approaches, promoting a better grasp of sustainability that combines responses to modern global issues.

B. Key Milestones in Environmental Policy and Practice

The changes in environmental policy and practice have seen big shifts, largely due to historical events that pushed for a reevaluation of how humans relate to nature. A key moment was the creation of the National Environmental Policy Act (NEPA) in 1969, which required thorough environmental checks for federal projects, making environmental issues a regular part of government decisions. These guidelines not only set a standard at home but also helped form international agreements, like the 1992 Earth Summit that sparked global talks on sustainable development. As knowledge increased about the complex connections between biodiversity, economic factors, and environmental health, discussions changed, urging stakeholders to adopt broader strategies—highlighting the connections between ecological, economic, and social aspects in policy-making. As a result, big initiatives like biodiversity offsets came about, aiming to reconcile development with environmental care (Abdo et al., 2019). This path shows a vital acknowledgment of the need for united approaches in building a sustainable future, supporting the idea that thorough governance is key to creating a strong ecological structure.

I.   Case Studies in Urban Sustainability

New challenges in urban sustainability show big gaps in how we understand local views and environmental effects. A study of Brussels Airport shows this, illustrating how sharing information can influence public understanding of environmental problems linked to flying. Even with more focus on greenhouse gas emissions from the aviation industry—which rapidly contributes to climate change—the local impact, especially on air quality and noise issues, often gets little attention ((Boussauw et al., 2019)). This way of communicating not only hides the real environmental impact of flying but also points out the differences in regulation among different sectors. These results highlight the need for a better approach to environmental communication that shows both the operational impacts and how the public sees them, making it vital in the search for sustainable urban areas ((Dewulf et al., 2017)).

A. Innovative Urban Planning: The Role of Green Architecture

The way cities look, and function is changing more because people see the need to be good for the environment and live sustainably. In this view, new ways of urban planning become very important, showing how buildings and caring for nature can work together. Green architecture, which focuses on using less energy, incorporating nature, and using sustainable materials, is not just about looking good; it's a key method to reduce the negative effects of cities on the environment. Examples show that cities that follow these ideas not only improve their environmental standings but also boost local economies by raising property values and getting communities involved. For example, projects like those from Transition UGent highlight how working together can help make sustainability a part of city planning, leading to real changes in urban design that fit with the goals of market growth (Block et al., 2016) (Madeira et al.). These changes offer a guide for future city planning, signaling an important move toward cities that can adapt and be strong.

B. Community Engagement in Sustainable City Initiatives

The growing difficulty of city problems needs new ideas that mix community involvement with sustainable city plans. By encouraging everyone to take part, cities can use different viewpoints to help create solutions that fit local needs and values. Involving residents often leads to more interest in local projects, which are key to being sustainable. For instance, projects like ACCESS promote the use of common reporting standards among nonprofits, improving clarity and responsibility while rallying local support for sustainability efforts (('Association for Vascular Access,' 2003)). Also, teaching programs focused on systems thinking, such as those in the STiP framework, help participants work together on tough issues, highlighting the need for a comprehensive view of city problems ((Blackmore et al., 2015)). In the end, strong community engagement not only builds social unity but also boosts resilience against environmental challenges, aligning with the main aims of sustainability.

II. Case Studies in Agricultural Sustainability

The complex idea of agricultural sustainability goes beyond just growing crops. It also includes keeping ecosystems balanced, making sure farming is economically sound, and ensuring fairness in society. Examples of new practices, like agroecology and permaculture, show how these methods can improve food security and heal ecosystems. For example, a significant project in agroforestry shows how planting trees in farming areas can support biodiversity and enhance soil health and carbon capture. These practices help lessen the effects of climate change by increasing strength against severe weather (Bansal et al., 2015). In addition, educational initiatives that promote systems thinking, like the Open University's postgraduate System Thinking and Practice (STiP) program, highlight the need for comprehensive strategies in agriculture (Blackmore et al., 2015). Together, these examples demonstrate that sustainable farming methods are crucial for building strong food systems, serving as guides for a greener future.

A. Permaculture Practices and Their Impact on Biodiversity

The link between ecological diversity and permaculture practices is very important. These methods push for a redesign of farming systems that are productive and ecological. By using strategies that copy natural ecosystems, permaculture creates habitats that support many species, thus increasing biodiversity. Studies show that these ecological practices improve soil health, save water, and cut down on the need for chemical inputs that hurt local ecosystems (cite20). Moreover, the digital era offers a chance to promote these practices with online resources, making information easier to find and expanding methods for permaculture (cite19). The combination of technology and caring for the environment shows a change towards sustainable farming, with different crops being grown and native species being kept that support a strong ecosystem for both people and the environment. This approach may help reduce biodiversity loss in a quickly changing climate, encouraging a better and more sustainable way of living together.

B. Technological Innovations in Sustainable Farming

The relationship between technology and sustainable farming shows a big change in how agriculture is done, combining care for the environment with smart economic choices. Various new methods, like precision agriculture, use data and Internet of Things (IoT) tools to boost crop production while reducing resource use. This industry responds actively to the challenges of climate change and food security. Additionally, adding agroecological ideas into these tech methods creates a complete approach where biodiversity and healthy ecosystems are just as important as high yields. For example, research projects discussed in the Implementation Action Plan support teamwork that includes various stakeholders, promoting inclusive innovation relevant to both organic and conventional farms (Cuoco et al., 2010). Such broad strategies not only help share knowledge but also encourage major progress in sustainable practices, which builds the community's strength in changing agricultural environments (Bellon et al., 2010).

For instance, Sweden has raised the standards significantly. Renewable sources now account for over 50% of the country's energy supply. This change was not sudden but rather the outcome of extensive planning and a strong dedication to decreasing reliance on fossil fuels. Sweden is showing that setting ambitious goals can lead to significant change by aiming to completely eliminate carbon emissions by 2045. Investing in electric buses, smart grids, and energy-efficient buildings is aiding in achieving these goals. Sweden has also adopted urban farming initiatives, converting empty spaces into productive green areas. These actions not only lower emissions but also improve urban life quality, promoting healthier surroundings and presenting fresh prospects for economic development.

Denmark offers another powerful example, particularly in its biggest urban area, Copenhagen. Copenhagen is well-known for its ambitious aim to reach carbon neutrality by 2025, with sustainability being integrated into every part of its urban development. The advanced biking system in the city encourages locals to abandon their vehicles, reducing air pollution and easing congestion. Prioritizing renewable energy and transportation, as well as Denmark's reliance on wind power for about half of its energy, illustrate how significant results can be achieved. Copenhagen is not just reducing its carbon emissions but also evolving into a more pleasant and breathable city that combines sustainability and convenience.

In Freiburg, Germany, the Vauban neighborhood serves as a sustainable living model led by the community. In this area, cars are hardly present as streets are made for walking and biking. In Vauban, houses not only fulfill their energy requirements but also produce extra energy using solar panels and efficient planning. The distinguishing factor of Vauban is its strong community feeling. The neighborhood was planned with contributions from local residents, who participated in deciding how their community should operate. This inclusive method ensures that sustainability is not imposed from above but is seen as a shared responsibility. The outcome is a community that values both environmental health and social well-being, demonstrating that sustainability can be centered on personal and community needs.

The case studies showcase a larger trend occurring globally: the incorporation of ecological well-being in economic strategies. Ideas such as the circular economy are becoming more popular, changing the way products are created, utilized, and repurposed. Rather than following the conventional "take-make-dispose" method, the circular economy prioritizes prolonging the use of resources. This model encourages redesigning products to allow for repair, reuse, or recycling instead of being disposed of in landfills. Reconsidering waste is leading industries to discover fresh techniques to lessen environmental harm, as well as generate employment and encourage creativity. This system demonstrates that sustainability and profitability can be mutually advantageous for the economy and the environment.

Simultaneously, green finance is aiding in unleashing the potential of these sustainable innovations. Green finance channels money toward projects with beneficial environmental impacts through loans, investments, and incentives. Green finance is essential for turning sustainability into a tangible outcome, whether by financing wind farm development, backing energy-efficient housing, or supporting emissions-reducing infrastructure projects. It's a strong tool that links financial institutions, governments, and businesses to ensure the availability of resources for supporting large environmental projects.

Another crucial component of the puzzle is to invest in ecosystem services. Forests, wetlands, and oceans are natural ecosystems that offer essential services such as clean water, air, and biodiversity to sustain life on Earth. These systems are essential for controlling the climate, preserving biodiversity, and supplying resources such as wood and seafood. By acknowledging the importance of these services, governments and businesses can make better decisions that prioritize the long-term health of the environment. One example is that preserving forests not only preserves biodiversity but also aids in storing carbon, which reduces the rate of climate change. Preserving these ecosystems has obvious economic advantages because flourishing environments contribute to thriving economies.

These instances demonstrate that sustainability is not just an abstract objective, but is being achieved in practical ways across the globe. Cities are currently adopting renewable energy and communities are altering their lifestyles and work habits in the current shift toward sustainability. It is important to not only minimize damage but also to build sustainable systems that promote the health of both humans and the environment.

The push for a sustainable future is gaining strength, fueled by communities, industries, and governments acknowledging the importance of balancing economic development with ecological stewardship. The future of sustainability involves rethinking how we live and making smarter choices that benefit everyone, as shown by these examples. Taking cues from countries such as Sweden, Denmark, and Vauban, the plan for a sustainable future becomes more evident, demonstrating that a world where both people and nature can thrive is not only possible but can be realized.

Eco-Innovators: Profiles of Progress

In the drive toward sustainability, a few individuals and businesses have emerged as key players, influencing the global discourse and setting a positive example. These eco-innovators are not only gaining attention in the media, but also putting into action practical solutions that combine sustainability with economic advancement. Their efforts, whether on a small scale or worldwide, show the power of innovation and environmental consciousness in creating lasting impact.

Elon Musk is one of the most notable figures in this movement. His collaboration with Tesla has transformed the car sector by introducing electric vehicles to the general public. Prior to Tesla, electric vehicles were frequently perceived as not realistic or catering to a specific audience. Musk transformed that view, turning electric cars into a representation of both extravagance and eco-friendliness. Tesla has made advancements in battery technology, including creating batteries that are more efficient and last longer, which have expanded the potential of electric cars. This not only reduced the price of storing renewable energy but also prompted other car manufacturers to speed up their transition to electric vehicles. Musk's vision goes further than just vehicles. By using Tesla's energy solutions like solar panels and the Powerwall battery, households can achieve self-sustainability by producing and storing renewable energy. This goes beyond just revolutionizing transportation - it's a comprehensive strategy for building a sustainable future in energy.

Patagonia exemplifies how companies can incorporate sustainability into their core values. Since the start, the outdoor clothing and gear company has consistently focused on promoting a strong pro-environment philosophy. In addition to selling products, Patagonia promotes its environmental agenda through various means. The company sets a high standard for waste reduction in manufacturing by using recycled materials. The 'Worn Wear' initiative promotes repairing Patagonia gear instead of buying new ones, extending their use and reducing the need for new materials. This challenges consumerism that prioritizes newness over durability, turning products into disposable waste. And Patagonia's commitment to activism goes beyond just its products. Each year, it gives millions of dollars from its profits to support grassroots environmental organizations, demonstrating that environmental protection and financial gains can go hand in hand. It indicates that the Earth can serve as a source of pride in a company's brand. It also demonstrates that loyalty can be cultivated through environmentalism. Consumers are embracing this trend and connecting more and more with companies that align with their values.

Through projects such as the purchase of green energy to power its operations across the globe, Google has set the pace for companies in the technology industry to become more sustainable. When Google announced that it would use only renewable energy to power all its operations worldwide, costs fell, and others in the technology industry began to see new opportunities. Google purchased energy from wind farms and solar installations, not only for its own energy requirements but also to boost the growth of renewable energy resources. Google used artificial intelligence to improve the efficiency of its energy-intensive data centers, helping to cool and regulate a multitude of inputs in order to reduce energy usage, which helped lower Google's carbon footprint and set the bar for the tech industry. Google shows that one does not have to scale back innovation or expansion in large firms to see large-scale environmental changes.

On a more individual basis, Greta Thunberg has emerged as the symbol of climate activism driven by young people. The initial solo school strike in Sweden has evolved into an international movement, motivating millions to demand immediate climate action by participating in street protests. Thunberg's straightforward approach to addressing the seriousness of the climate crisis has compelled political leaders to face the truth about environmental damage. Her advocacy has not just increased awareness but also altered the global conversation, elevating climate change to a key topic in international policy talks. Greta's influence goes beyond just politics; she has proven to people everywhere, especially the youth, that their opinions count and that working together can result in actual progress.

These innovators have adopted a fresh perspective on the connection between economic development and environmental protection. Companies such as Tesla and Patagonia show that advancing products and business strategies can increase profits while also minimizing environmental damage. Google's emphasis on renewable energy and energy efficiency shows how big operations can reduce their environmental impact with intelligent investments. Through her activism, Greta Thunberg highlights that sustainability should be a personal responsibility, not just a corporate one.

The common thread among these innovators is their commitment to ensuring their sustainability over the long term. It is evident now that change does not always have to be radical. Thinking in new ways about how raw materials are used, product design or business operations can lead to many accomplishments. Their behaviors result in significant impacts, such as reducing carbon emissions and protecting natural resources. Their powerful message is that global issues can be effectively addressed through creativity.

The lesson from this creativity and the wider learning of eco-innovators is that sustainability thinking should be ingrained in our everyday lives. Technological advancements, innovative ways of doing business, and community-based advocacy provide a preview of the future in which economic development and environmental well-being go hand in hand. Stories like these also elevate anticipation by demonstrating that adopting sustainability thinking can steer us toward a new mindset if we are open to it.

Cultural Shifts Toward Sustainability

Our cultural connection to nature and resources is evolving, not solely due to heightened awareness of environmental issues. These changes indicate a stronger link between human health and the health of our surroundings. Influenced by a shift toward simplicity, conservation of the environment and better management of biodiversity, we are already altering our lifestyles, our economies, and our policies.

Arguably, minimalism is the pivotal movement driving this change. The concept of minimalism generally involves leading a more basic life, possessing fewer items and concentrating on what is truly important. However, it is now more than just that. Minimalism, particularly in relation to purchasing goods, involves carefully thinking about the quantity of items we buy and promoting the decision to purchase fewer items of higher quality. Essentially, this involves having fewer yet better quality possessions that have a longer lifespan, thus decreasing waste and conserving resources. In a time of excessive consumerism, minimalism provides a sharp contrast. Minimalism focuses on living a more environmentally friendly life by highlighting simplicity and sustainability. It involves having fewer possessions but selecting items that will cause minimal harm to the environment.

The eco-minimalism trend has taken this a step further, focusing on minimizing environmental impact. For instance, adherence to a zero-waste lifestyle is gaining traction as individuals actively steer clear of disposable plastics, repurpose or recycle biodegradable waste, and select alternative eco-friendly materials. The rise of compact, eco-friendly houses like tiny homes indicates that more individuals are interested in adopting a sustainable approach to consumption. Eco-minimalism aims to encourage people to consider the footprint of their accumulation – shifting their habits toward keeping and hoping to expand people's understanding of their impact on the environment. The sustainability movement is also getting stronger as increasing numbers of people make green choices, from urban dwellers adjusting their lifestyle to living in smaller homes, eating more sustainably, and using public transport more often.

This cultural shift makes people see the environment in a different light and perhaps consider the value of habitats and animals – especially endangered ones – more highly than before. In many conservation projects, re-establishing an ecosystem to a state resembling a previous status is an important objective. It is a proactive method of restoring nature and can reestablish crucial biodiversity in areas where it has disappeared. While protecting land and conserving species are important, rewilding takes it a step further by actively restoring them. Rewilding Europe is currently a top illustration of this in Western Europe, focusing on initiatives to rewild and revive populations of dwindling species like wolves and bison in their natural environments. Rewilding not only benefits ecosystems but also has a favorable effect on communities and tourism. It has the potential to generate fresh possibilities for ecological tourism, attracting financial resources to distant wilderness regions.

At the same time, the focus remains on preserving undisturbed ecosystems such as rainforests, coral reefs, wetlands, and other natural habitats. These are the diverse, life-sustaining, carbon-rich, ecosystem-maintaining, and supporting characteristics of Earth's surface that are crucial to preserve as our planet faces climate change. We desire for these undisturbed ecosystems - the final vast areas of untouched natural environment - to remain unchanged. This concerns services currently taking place in locations like Costa Rica, with a focus on national conservation efforts. The nation safeguards large areas of forested land and carries out programs to plant new trees. Costa Rica exemplifies sustainable practices by successfully conserving biodiversity and decreasing deforestation without causing a rise in poverty. Conservation has been embraced as both a national policy and a cultural value, and it has proven to be effective.

A growing respect for biodiversity signifies a shift in cultural awareness regarding the importance of ecosystems to human society. Ecosystems with high biodiversity are not only visually appealing but also essential for supporting life. They offer vital services like pollination, pest control, and climate regulation. The recent recognition is starting to impact changes in economic systems toward actions that support the conservation of biodiversity. For instance, more people are viewing the idea of a circular economy as a method to create products that minimize waste and repurpose resources. Fixing, repurposing, and recycling help lessen the harm to the environment from extracting natural resources and also encourage innovation. In this development model, industries are urged to consider resource flows in the long run, decreasing resource consumption to allow for future flows and opening up economic opportunities for new activities.

A sustainable agenda greatly relies on green finance as a key component. In conjunction with the circular economy, it is more and more bridging the void created by the unrestricted development approach. Green finance involves investing in projects or services that benefit the environment, like sustainable agriculture, eco-friendly infrastructure, or renewable energy. The funding of green finance is increasingly supporting projects focused on decreasing emissions, utilizing resources more effectively, and protecting ecosystems. Due to the growing green finance sector, economic goals are now more focused on environmental objectives, making sustainability both achievable and lucrative. Investing in projects that support environmental well-being is crucial in transitioning to a more eco-friendly economy, making green finance a vital component.

Japan demonstrates how cultural changes can support the country's sustainability goals. Japan's historical focus on pollution control has expanded to include climate mitigation, resource efficiency, and biodiversity conservation. The nation has launched effective waste minimization movements, incorporating recycling initiatives into everyday routines. Cities such as Tokyo have adopted green infrastructure, incorporating rooftop gardens and energy-efficient buildings. These endeavors demonstrate how powerful cultural beliefs, such as reverence for nature and community, can lead to notable environmental advancements. Japan's strategy shows that sustainability can be integrated into society through education, policy, and cultural awareness.

While these cultural changes keep growing in popularity, they show a preview of a potential future where sustainability is a given. Societies can develop resilient and adaptive systems by following minimalism, promoting conservation, and prioritizing biodiversity. These changes are not only focused on protecting the environment, but also on guaranteeing the welfare and success of people. As more people and communities understand the link between ecological health and economic growth, the opportunity for working together increases, leading to a sustainable and flourishing planet.

Sustainable Governance: Shaping Climate Policies

Recently, there has been a change in the political atmosphere where addressing climate change by incorporating low carbon issues into mainstream policy agendas is now considered crucial. This shift is driven by the understanding that tackling climate change requires a more integrated approach that considers economic, social, and ecological goals.

A significant achievement was the Paris Agreement, finalized in 2015, where 195 nations agreed to restrict global warming to under 2°C, preferably below 1.5°C. This marked a unique approach where countries set their own climate targets, called Nationally Determined Contributions (NDCs). Unlike prior treaties like the Kyoto Protocol that imposed strict goals on developed countries, the Paris Agreement is more flexible, urging all nations to get involved based on their individual situations and capabilities. Sweden, Norway, and New Zealand have committed to reaching net-zero emissions by 2050, setting an example for other nations to emulate.

In addition to documentation, this framing is now also being inscribed into the very fabric of economic and environmental planning. The idea of the circular economy now has widespread buy-in and is backed by powerful multi-stakeholder groups that advocate for higher levels of resource use efficiency and reduced waste, all while economic growth occurs. In contrast with the tried-and-tested (but wasteful) 'take-make-dispose' model that underpins many industrial processes, the circular economy emphasizes the creation and use of products that can be reused, repaired or recycled. Lower pressure on natural resources and with it, new economic opportunities would be the result. At the same time, green finance is backing these ambitions by steering capital to projects that promote sustainability, such as renewable energy, sustainable agriculture, and green infrastructure.

Nations are demonstrating how a proper combination of strategies can revamp their economies while tackling environmental issues. China has quickly grown its solar power sector and is now the top global manufacturer of solar panels. This change has not just decreased emissions but also generated employment and lowered the cost of solar power worldwide. Mexico has adopted waste-to-energy technologies, converting waste into a valuable resource and decreasing its dependence on fossil fuels. These instances demonstrate how governments utilize economic strategies to reach environmental objectives, showing that sustainability can spur innovation and development.

Successfully governing requires carefully balancing economic priorities with climate goals. Numerous nations have discovered that tackling climate change not only helps decrease air pollution but also enhances energy security. India's ambitious solar power initiative serves as a prime example, as it has assisted in reducing the country's reliance on coal while offering cleaner energy to millions of individuals. Likewise, Colombia is incorporating eco-friendly practices into its construction industry, demonstrating how sustainability can be seamlessly merged with conventional sectors to promote growth. These nations are not simply responding to climate change; they are establishing a future in which sustainability is integrated into the economic bedrock.

Nevertheless, there are obstacles despite these favorable advancements. Challenges such as political resistance, economic limitations, and social disparities may hinder the advancement of climate policies. Certain governments find it challenging to afford the shift to environmentally friendly energy sources, while others encounter resistance from sectors linked to non-renewable sources. In order to address these challenges, countries are trying out tactics such as providing social benefits to at-risk groups impacted by the move to eco-friendly jobs or promoting climate regulations when oil prices are low to lessen the impact of the transition. The goal of these methods is to create a more equitable and acceptable shift to a green economy, ensuring no one is excluded during the transition to sustainability.

The progression of political strategies on climate change demonstrates an increasing awareness that effective governance necessitates a comprehensive and all-encompassing structure. Countries must do more than just reduce emissions; they must also ensure that their economic growth is in line with protecting the environment. As more countries begin using this method, they are setting an example for others to emulate, demonstrating that it is feasible to address climate change and simultaneously strengthen the economy. As countries refine their policies and work together on global solutions, the challenging path ahead shows potential for clearer progress.

Climate Jobs & Opportunities

Climate Jobs

Today, the connection between jobs and climate action is getting more attention. It is clear that big changes in the workforce are needed to fight climate change, making the idea of climate jobs very important. Climate jobs refer to various roles that help the environment and strengthen economic stability in areas like renewable energy, green transportation, and eco-friendly construction. These jobs are very important because they are essential for making the systemic changes needed to tackle the complex challenges of climate change. Also, building a skilled workforce is crucial to meet climate targets, which requires specific educational programs and thorough training. As society faces the urgent need for ways to reduce climate impacts, having a well-trained workforce will be key to moving toward a sustainable economy.

A. Definition of climate jobs and their significance in mitigating climate change

The move to a sustainable economy needs a clear understanding of climate jobs. These jobs focus on cutting down greenhouse gas emissions and helping people adapt to climate change. They cover various areas like renewable energy, green transportation, and eco-friendly construction, showing how work and care for the environment are closely linked. Climate jobs are important not just for reducing climate change's impact but also for building economic strength by creating lasting job opportunities. As the need for clean energy solutions and green practices grows, there is an increasing demand for skilled workers who have the right knowledge to support this shift. Better workforce training programs are crucial to meet these needs, making sure people are ready to play a key role in the growing climate economy and achieve larger climate goals (see (Aldy et al.) and (Joseph E. Aldy et al.)).

B. Overview of sectors contributing to climate resilience

A broad method is important for understanding how different sectors help climate resilience, showing a complex connection between economic activities and ecological sustainability. For example, the renewable energy sector not only creates jobs but also plays a key role in lowering greenhouse gas emissions, leading to a shift toward sustainable energy systems. The transportation sector is also significant, as it is moving toward sustainable methods, like making electric vehicles and improving public transit, which generates new job options while reducing carbon footprints. Forestry and conservation efforts highlight the need to maintain biodiversity and ecosystem health, fostering jobs in restoration and management. As shown in (UNDP, 2016), combining climate action with sustainable development is key to building resilience, supporting the claim that developing the workforce in these sectors is vital for advancing climate goals and ensuring sustainable livelihoods.

C. Importance of workforce development in achieving climate goals

The growing field of climate jobs includes an important link between workforce development and eco-friendly sustainability, particularly in building and construction. As green building methods become more popular, there is a rising need for skilled workers in energy-saving retrofitting and sustainable construction, highlighting the need for strong workforce development programs. These programs train workers in necessary skills and help them shift to environmentally friendly building methods that support climate goals. By focusing on detailed training that covers green certifications, safety procedures, and new construction techniques, interested parties can build a workforce ready to handle the challenges of sustainable practices. Moreover, these efforts lead to significant social and economic benefits, helping create jobs while building resistance to climate change. Therefore, building a well-trained workforce is not just a practical issue; it is essential for reaching larger climate goals (Dolf Gielen et al., 2019) (2018).

D. Historical context of climate jobs and workforce evolution

The growth of climate jobs has been heavily influenced by past events and the slow rise of environmental awareness in job development. The Rockefeller Foundation's Sustainable Employment in a Green US Economy (SEGUE) initiative, which started in 2009, shows an important effort to connect green jobs with fair socioeconomic opportunities, mainly aiming at low- and moderate-income workers. This is done through funding in green sectors like energy efficiency and waste management (Carlos Martín et al., 2013). As climate change has become more urgent, the need for a skilled workforce has become more obvious, leading to programs aimed at job creation along with targeted training (Surdna Foundation, 2009). This link between jobs and environmental improvement shows how workforce growth not only reacts to new technology in renewable energy but also tackles old inequalities, creating a key route for a strong economy based on sustainable methods.

E. Objectives and scope of the research essay

The study of Climate Jobs needs a clear framework that defines its goals and scope in sustainable economic development. This research essay aims to explain the complex nature of climate jobs in various sectors, especially in renewable energy, agriculture, transportation, and conservation. The analysis will review how workforce development programs can align with new job opportunities in these areas, helping to build climate resilience and support a fair transition for workers affected by industrial shifts. By examining examples like the Rockefeller Foundation's Sustainable Employment in a Green US Economy, the research will highlight the need to invest in specific training programs that encourage sustainable practices and skill development across different industries (Carlos Martín et al., 2013). In addition, the study will look at current obstacles and suggest practical policy reforms to prepare the workforce for a resilient economy, ultimately helping to reduce the effects of climate change.

II. Renewable Energy Sector

The renewable energy sector has a big chance to change things, going beyond just cutting emissions; it shows a new way of building economic systems that focus on creating jobs and being ready for climate challenges. As countries decide to move toward low-carbon methods, there's been a big rise in the need for workers with skills in renewable technologies, especially in solar and wind. This has led to a need for workforce development efforts. For example the global solar photovoltaic (PV) industry is a good example of this, as it helps not only with manufacturing jobs but also with roles in research, project development, and installation (Jacob Funk Kirkegaard et al.). Because of this growth, climate policies should focus on training programs that give workers the skills they need, making sure they are ready for the changing job market in this fast-evolving area (Młynarski et al., 2015). Thus, putting money into training people in the renewable energy sector is key for building a strong economy that can handle both climate change and the rising job issues it brings.

A. Employment trends in solar, wind, and other renewable energies

The growing need to deal with climate change has led to a big rise in job chances in renewable energy fields, especially in solar and wind. This increase shows a wider change in society toward sustainable actions and an understanding of how important workforce training is for meeting climate objectives. Specifically, reports say about $36 trillion is needed for clean energy investments by 2050, which points out the job opportunities that could come with expanding renewable energy systems (Mark Fulton et al., 2014). Also, strong efforts to create labor-driven projects, like those in New York, show how effective it is to link climate actions with economic fairness (Cha et al., 2017). These mixed strategies not only support a shift to low-carbon economies but also stress the need for specific training programs that give workers the right skills for these changing jobs, making sure the workforce is ready for future energy needs.

B. Skills required for renewable energy jobs

The quickly changing field of renewable energy needs different skills to have a qualified workforce to face its issues. Skills in engineering, especially in electrical and mechanical systems, are very important for designing and improving renewable energy technologies, like solar panels and wind turbines. Also, knowledge in data analytics and software programming is becoming more important, as advanced technologies and grid integration require better data management and predictive maintenance (cite15). In addition to these technical abilities, good project management skills help in effectively carrying out complicated renewable energy projects, making sure that timelines and budgets are followed. Soft skills, such as communication and teamwork, are important since working together in diverse teams is crucial for the success of projects. As the world economy moves towards sustainability, developing these skills will be vital in building a capable workforce ready to handle the demands of the renewable energy sector (cite16).

C. Case studies of successful workforce training programs

Creating new workforce training programs is very important for helping marginalized communities and increasing their roles in climate jobs. Existing programs show how focused training can build skills needed to handle social and economic issues alongside climate challenges. For example, the Green Jobs Movement focuses on making job opportunities for low-income individuals in sustainable fields, helping to solve problems related to economic inequality and environmental harm ((Green for All, 2010)). Additionally, frameworks like the Pathway to Successful Young Adulthood promote strategic, cross-discipline actions that improve community conditions, aiding young people's shift into green jobs while achieving climate objectives ((Lisbeth B. Schorr et al., 2007)). These insights highlight the need to blend workforce training with wider socio-economic plans, demonstrating that effective programs not only provide vital skills but also lead to better community health and resilience, promoting a strong and fair workforce for the future.

D. Economic impact of renewable energy job creation

Changing to an economy based on renewable energy offers many chances for job growth, which can boost economic progress and fairness in society. Research from E2 shows that there are many job opportunities from the faster use of clean energy technologies, supporting the idea that we can fight climate change while reducing economic inequality (Rebecca Deehr, 2016). Additionally, examples from New York State show how organized labor can significantly influence climate policies that address both joblessness and environmental issues. These initiatives demonstrate that a well-structured climate jobs plan, which centers on sustainable jobs in areas like solar and wind energy, can create good jobs that support families (Cha et al., 2017). Therefore, developing the workforce strategically, focusing on training and gaining skills, is crucial for maximizing the economic benefits of renewable energy and building a strong and fair economy.

E. Challenges faced by the renewable energy workforce

Switching to a renewable energy economy involves a difficult situation for workers, with big problems that can stop progress and growth. Major obstacles include a widespread skills gap, as many current workers do not have the necessary technical skills for new technologies in solar and wind energy (2008). Also, the fast pace of tech changes can leave training programs unable to keep up with what the industry needs, worsening the issues in the workforce. Moreover, areas that depend on traditional energy industries face economic challenges when jobs in fossil fuel sectors disappear, putting local economies and tax income at risk, which makes communities vulnerable during this change (Hamilton et al., 2017). Tackling these problems requires focused work on policy and education reform to help workers who lose jobs move into sustainable energy jobs, thus supporting strength and adaptability in the renewable energy field.

III. Transportation

The shift to sustainable transportation systems is changing jobs in many areas. The move to electric vehicles (EVs) requires a large workforce to make, fix, and use these new technologies, leading to the rise of specific jobs for this new market (Hoppe et al., 2015). Also, public transit systems, which often lack funding, are being updated to handle more riders and cut down on carbon emissions. This creates many green jobs, from operations to building projects (Hlalele et al., 2012). To prepare the workforce for these changes, special training programs are now available to help workers gain the needed skills for green transportation jobs. These programs not only improve job chances but also help ensure a fair transition so that people who have depended on carbon-heavy jobs can benefit from climate-related employment. Overall, adding environmental issues into transportation plans shows how important this sector is for building a strong and sustainable economy.

A. Transition to sustainable transport systems and its impact on employment

The move towards better transportation that is good for the environment brings not just benefits for nature but also important job chances as communities work to fight climate change effects. Different studies have shown that this change involves new, specialized jobs in the electric vehicle industry and improvements in public transport, fitting with plans to lower greenhouse gas emissions in cities. Projects like the Carbon Free Boston initiative show how crucial it is to make sure these changes are fair, stressing that job training should focus on communities that have been left out in the past to ensure everyone can get in on the green jobs (Boston University Institute for Sustainable Energy et al., 2019). Additionally, changing laws to support green economies is very important, as shown in the Rio 20 Declaration, which calls for new legal ideas to manage the job issues that come with these changes (Gehring et al., 2013). Therefore, redoing transportation systems is key to helping the environment and creating a strong and fair job market.

B. Emerging jobs in electric vehicle production and public transit

The move toward a sustainable transportation system is very important for dealing with the climate crisis. This creates many new job chances, especially in making electric vehicles (EVs) and improving public transport. As cities and governments focus more on low-emission transport to reduce greenhouse gas (GHG) emissions, the need for skilled workers in the EV field is growing. These jobs not only include regular manufacturing roles but also involve software and technology development, battery making, and maintaining infrastructure, which all require a mix of skills (The Pew Charitable Trusts, 2009). In addition, public transit systems are changing a lot to improve access and efficiency, while also meeting carbon neutrality goals. This is seen in plans like the Carbon Free Boston report, which focuses on electrification and fair access (Boston University Institute for Sustainable Energy et al., 2019). Therefore, specific training programs are crucial to prepare the workforce with the skills needed to succeed in this changing job market, helping to ensure a fair and sustainable economic transition.

C. Training initiatives for green transportation jobs

As the world moves towards sustainable transport systems, the need for skilled workers in green transport jobs grows more important. Initiatives from governments and the private sector are key in creating training programs that help people gain the skills they need for this changing job market. These programs focus on electric vehicle production, including making and fixing vehicles, and on improving public transit systems with green practices. By adding green values to transportation education, programs can meet the rising need for technical skills that support clean technologies, like electric and hydrogen vehicles. Helpful partnerships between public and private sectors can further improve job chances and help build a workforce ready to lead change in urban transportation solutions ((T. BUIH. et al., 2019)). In the end, strong training frameworks will boost economic growth and lead to important progress in climate resilience ((Tofikk Redi, 2024)).

D. Role of policy in shaping transportation job markets

Job markets in transportation change a lot because of policies focusing on sustainability and climate resilience. As countries shift to electric vehicles (EVs) and better public transport, government actions are key in affecting jobs in these areas. For example, policies that promote EV adoption help increase consumer interest and also create jobs in manufacturing, upkeep, and building related to these technologies. Additionally, training programs aimed at green transportation jobs make sure workers have the necessary skills for new market demands, helping to lower employment barriers for underrepresented groups, such as people with disabilities who face special issues with access and costs (Katie Savin et al., 2024). This mix of policy and job growth forms a sustainable employment scene, underlining the importance of strategic efforts to create a strong climate job market in transportation (Hemalatha J. et al., 2024).

E. Future trends in sustainable transportation employment

The global focus on sustainable economic practices is making changes in job chances in transportation. The move to sustainable transport, which includes more electric cars and better public transit systems, is expected to create many new jobs like battery tech experts and smart transit systems analysts. To get workers ready for these new jobs, specific training programs will be important to teach skills needed for green technologies and sustainable methods. Furthermore, as cities focus more on eco-friendly systems, legal rules that support this shift will be important to handle the rules governments must follow. As mentioned in (Gehring et al., 2013), these changes will not just help build a greener economy but also create many job opportunities in sustainable transportation, greatly affecting climate job trends.

IV. Building and Construction

The mix of new building methods and climate readiness is changing fast in the construction industry, creating big chances for economic growth through jobs related to climate. Current green building projects are not just about using sustainable materials; they also focus on improving energy efficiency, which is important for cutting down total carbon emissions. Studies suggest that applying these methods can create jobs at many levels, ranging from skilled workers to project managers focused on sustainability (Sayali Andhare et al., 2024). In addition, strong training programs in green construction are becoming very important for teaching workers the skills they need, making sure they can fill new positions that focus on ecological issues (Ivan Bozhikin, 2023). By promoting a job market that values eco-friendly methods, the construction industry can help local economies and contribute to global climate objectives, leading to a stronger economy.

A. Green building practices and their influence on job creation

The growth of jobs in green building practices is more and more seen as important for a sustainable future. With the need for energy-saving and eco-friendly buildings increasing, people in construction are matching their skills with green architecture principles, which often involve updating old buildings and using new material technologies. Programs like the Rockefeller Foundations SEGUE have shown the importance of workforce training in this area, concentrating on helping low- and moderate-income workers join green sectors, thus tackling both job and environmental needs (Carlos Martín et al., 2013). Additionally, by adding sustainability to building methods, these programs not only help the economy grow but also support health fairness by lessening exposure to environmental risks, especially in underserved areas (Green for All, 2010). The close link between green building practices and job creation shows the wider potential of climate jobs to build strong economies.

B. Roles in energy-efficient retrofitting and sustainable construction

The need for sustainable practices in the building industry creates a growing link between worker training and energy-saving renovations. As the world becomes more aware of climate change, energy-efficient retrofitting is important for cutting carbon emissions in older buildings and offers many job chances. Workers in this area need skills in new building materials, energy checking, and smart design methods, highlighting the need for strong training programs (cite39). Additionally, sustainable building methods require teamwork from different parties, such as architects, engineers, and lawmakers, to make sure that both new projects and renovations reach top energy efficiency and durability (cite40). Combining these roles not only boosts economic development but also aids the shift to low-carbon cities, helping to meet larger climate goals. Therefore, training a workforce skilled in these areas is crucial for unlocking the potential of climate jobs in the building industry.

C. Workforce development programs in green construction

In a changing economy that is more focused on climate resilience, starting workforce development programs in green construction becomes an important way to deal with both job creation and sustainability. These programs help people learn skills related to energy-efficient retrofitting and sustainable building practices. They also work to connect traditional construction jobs with new demands from a green economy. Research shows that these initiatives can create a range of job opportunities, which are especially important in areas that have depended on fossil fuels, making sure workers can transition fairly to these changes (Cha et al., 2017). Additionally, by using inclusive training methods, workforce development in green construction can tap into the potential of people with disabilities, which improves workforce diversity and participation in climate-related jobs (Bruyere et al., 2013). This approach is crucial for developing a skilled labor force ready to handle the needs of a sustainable future.

D. Economic benefits of sustainable building practices

The link between sustainable building methods and economic growth shows great promise for creating jobs while addressing climate change. As construction uses more energy-efficient technologies and eco-friendly materials, new chances arise in different areas, especially for skilled workers in green retrofitting and sustainable building projects. These methods help cut down emissions and save money for building owners, which supports economic growth (Green for All, 2010). Additionally, when communities adopt sustainable development principles, improved job training programs can develop a skilled workforce ready to handle the changing needs of this growing field, creating a fairer economy (Green for All, 2011). By focusing on investments in sustainable building technologies, communities can tackle climate goals while boosting local economies, generating lasting job opportunities, and fostering a more resilient and sustainable future. This combined approach to economic and environmental health is vital for reaching larger climate objectives.

E. Challenges in implementing green construction initiatives

The building industry has major problems when it comes to using green construction habits that stop wider use. One big issue is the high upfront costs linked to sustainable materials and technologies, which can scare off those who don't know about the long-term savings. Moreover, the workers often do not have the necessary skills to use eco-friendly building methods, creating a need for specific training programs. Also, rules and regulations may not be well-developed or may not be applied consistently, making it hard for construction companies to follow green standards ((Carlos Martín et al., 2013)). Additionally, a lack of teamwork between private investors and construction firms can slow progress on eco-friendly practices ((United Nations Global Compact, 2015)). By overcoming these challenges, there is a chance to create more jobs in green construction, helping to move towards a stronger, climate-aware economy.

V. Energy Transmission, Distribution, and Storage

As societies move to a future with sustainable energy, updating energy transmission, distribution, and storage systems becomes crucial for both environmental gains and economic growth. Putting money into these areas is key, as it helps blend renewable energy sources into current grids and makes energy infrastructure stronger against climate-related issues. Jobs in this field require a wide range of skills; workers need to be skilled in classic electrical engineering along with new technologies used in smart grids and energy storage. The growing need for skilled workers in this field serves two goals: building a strong workforce to support and develop energy systems while also tackling major challenges from climate change (2008). As these systems change, the job market in energy is likely to grow greatly, creating a more sustainable and job-filled future in the green economy (Sustainable Development Commission, 2012).

A. Employment in grid modernization and energy storage solutions

The move to a lower-carbon economy is increasingly dependent on updates to the power grid and energy storage options, which are showing a key change in the energy field. As communities work to boost energy efficiency and strength, there is a growing need for skilled workers in these areas. Job openings include various roles, such as engineers, technicians, and project managers who focus on renewable energy systems and battery technology. With major federal funding, as mentioned in the American Recovery and Reinvestment Act, nearly $41 billion is set aside for energy projects that promote innovation and job growth in these sectors (Jason Walsh et al.). Additionally, workforce development programs need to focus on the technical skills needed to keep energy infrastructure strong, making sure that new workers are ready to deal with the challenges of modern energy systems. In the end, creating solid job pathways in these fields can play a big role in climate resilience and sustainable economic growth (Anastasia Christman et al., 2012).

B. Skills needed for maintaining resilient energy infrastructure

Moving to a strong energy system needs a mix of skills that cover technical, analytical, and personal areas. Workers in this industry must know modern technologies like smart grid systems, which need a good grasp of how hardware and software work together to improve energy distribution (cite51). Also, being skilled in data analysis is very important, as this helps with watching and predicting energy use patterns in real time to boost efficiency and responsiveness (cite52). Plus, good personal skills are key for working together with everyone involved, from engineers to policymakers, to make sure new solutions tackle social and economic differences in communities. As cities such as Boston aim for carbon neutrality, grasping the effects on equity is vital, which shows the need for training programs that foster inclusivity and adaptability in developing the workforce. So, giving people these varied skills not only strengthens energy systems but also helps create a fair shift toward sustainable energy.

C. Future job prospects in energy transmission and storage

The change in global energy systems is likely to create many job chances, especially in energy transmission and storage. As countries set big climate goals, investments in smart grid tech and better energy storage have increased, showing the need for a dependable and strong energy system. The International Energy Agency (IEA) estimates that to reach the needed decarbonization goal, an extra investment of about $36 trillion in clean energy will be needed by 2050, indicating a high demand for skilled workers in this field (Mark Fulton et al., 2014). Also, new developments in renewable energy technologies are leading to more jobs related to energy efficiency and management, as job growth is connected to moving away from fossil fuels (Christopher Flavin, 2008). So, developing focused workforce programs to help gain skills in these areas will be vital to take advantage of these new job chances.

D. Role of technology in transforming energy jobs

As industries use more advanced technologies, the change in energy jobs has become an important part of workforce growth for climate resilience. The use of digital tools, such as artificial intelligence (AI) and automation, greatly improves efficiency and safety in areas like construction and renewable energy. For example, safety technologies have changed methods in risky situations, boosting worker involvement and lowering injury rates ((Pravin Tathod et al., 2024)). Furthermore, the oil and gas industries' move to cleaner energy shows how vital technological innovation is for reducing greenhouse gas emissions, with methods such as carbon capture and better drilling practices allowing for more sustainable operations ((Ifeanyi Onyedika Ekemezie et al., 2024)). As a result, these technological changes require workers to have new skills, stressing the immediate need for specific training and educational programs that meet the changing needs of the energy sector. Therefore, technology reshapes job roles and supports the move to a sustainable economy.

E. Policy implications for Energy Workforce Development

As the need for skilled workers in climate jobs grows, it is important to make sure educational policies match the changing energy job market. An effective way to develop the energy workforce must consider the complexities of fairness and sustainability. It is essential to allow different communities to join this change for both social fairness and environmental effectiveness. For instance, adding fairness factors into job training programs not only tackles past inequalities but also strengthens community defense against climate challenges (Boston University Institute for Sustainable Energy et al., 2019). Also, the link between poverty and health highlights how skilled green jobs can improve overall health in communities, thereby having a healthier workforce (Green for All, 2010). By creating focused efforts that promote training in renewable energy and sustainable practices, policymakers can use energy workforce development to drive economic growth that includes everyone while also meeting climate goals.

VI. Conclusion

As the urgent issues of climate change threaten global stability, the link between climate jobs and workforce development stands out as important for a strong economy. The analysis shown in this essay points to a key conclusion: investing in various areas, from renewable energy to sustainable farming, is necessary to fairly reach climate goals. It is important to focus on inclusivity because disadvantaged communities often face the worst impacts of environmental harm and economic hardship. Studies show that good policies can create jobs while also tackling social inequalities ((Boston University Institute for Sustainable Energy et al., 2019), (Aquilina et al., 2011)). The future depends on teamwork among education, industry, and government to build a skilled workforce suited to the needs of a green economy. By taking this chance, society can reduce climate effects and build a strong job market that values environmental health and social fairness, ensuring a sustainable future for everyone.

A. Summary of key findings

A thorough look at how climate jobs and workforce development connect shows important findings that highlight how education and policy are key to building a strong economy. The Rockefeller Foundation's SEGUE initiative points out the need for doing sustainable jobs in green sectors, especially in energy retrofitting and water infrastructure, which have become crucial as green job markets grow (Carlos Martín et al., 2013). Also, the work environment affects how well employees perform across different age groups, indicating that customized workforce programs are needed to boost productivity in climate jobs (Tan et al., 2013). Additionally, workforce development programs should tackle skill gaps in new areas like renewable energy and sustainable agriculture, where focused training can improve worker readiness. Together, these findings support a planned effort to add climate-focused courses in schools, preparing future generations with the skills they need to succeed in the green economy.

B. The future outlook of climate jobs and workforce development

The changing landscape of climate jobs shows a big move towards a more sustainable economy, requiring strong workforce development plans. As states take on ambitious climate goals due to federal delays, there will be a noticeable rise in demand for skilled workers in renewable energy, sustainable transport, and green building sectors (Cha et al., 2017). These new job areas are key in fighting climate change and also offer a chance to help struggling communities that depend on fossil fuel jobs, highlighting the need for fair transition plans (Cha et al., 2017). Moreover, good education and training programs are crucial to give workers the STEM skills they need, making sure they can adjust to changing job needs (Keen Point Consulting et al., 2016). In the future, working together among government, businesses, and schools will be important in creating an inclusive workforce development strategy, ultimately helping to build a strong economy that meets global climate aims.

C. Call to action for stakeholders in education, industry, and government

Getting stakeholders from education, industry, and government to work together is very important for building a workforce ready for a climate-resilient economy. Schools need to change their curriculums to include sustainable practices and new technologies so students gain the skills needed for growing green sectors. At the same time, companies should develop apprenticeship programs and team up with schools to help graduates transition into good climate jobs. Government policies should encourage the creation of sustainable jobs through supportive laws and funding for training projects that focus on how climate affects health and infrastructure. As mentioned in talks about institutional investments, good teamwork can help clear up language differences and boost cooperation across different sectors, leading to a more combined effort for sustainable development ((United Nations Global Compact, 2015)). Working together from all parties not only tackles the urgent issues related to climate change but also helps build a strong economy where resilience and innovation can flourish ((Charles et al., 2010)).

D. Recommendations for enhancing climate job opportunities

As countries deal with the problems of climate change and economic recovery, it is very important to create climate job opportunities. Key suggestions include forming partnerships between schools and businesses to create specific training programs that fill skill gaps in the renewable energy, sustainable transportation, and green construction fields. By focusing on various disciplines, we can help close the skills gap, making sure that graduates can find different jobs in new markets. Policy measures like subsidies or tax breaks for companies that invest in workforce training can help boost job growth (Gretchen Kosarko et al., 2011). In addition, adding climate education to current curricula, along with vocational training, is crucial to preparing the next group of climate experts (Andrés Cisneros-Montemayor et al., 2017). These suggestions are connected to sustainable growth strategies, which not only generate job opportunities but also help build resilience against future environmental issues.

E. Final thoughts on the importance of a sustainable workforce for climate resilience

As world problems grow, having a sustainable workforce is very important for dealing with climate change. Moving toward a green economy needs workers who are not just skilled but can also adjust to new ways of handling environmental issues. A trained workforce can lead new methods in areas such as renewable energy and sustainable farming, helping to lessen climate change effects. Putting money into education and job training for new green jobs not only helps with current job needs but also builds lasting strength in communities. It is important to form partnerships between schools, industry leaders, and policymakers to create a strong plan for developing climate jobs. In the end, a sustainable workforce is key to a strong economy that can handle climate emergencies while supporting fairness and protecting the environment.

With more environmental problems and a strong need for new solutions, the goal of sustainability has become very important for societies everywhere. A varied approach that includes technology, policy, and community involvement is necessary to deal with the complex connections of ecological, economic, and social aspects involved in sustainability. This chapter aims to explain different case studies that show effective methods and practices to promote a greener future. By looking into successful sustainable projects, the analysis points out both the successes and the challenges faced along the way. These stories aim to motivate and inform policymakers, researchers, and practitioners, helping them better understand how working together can change our future. In the end, this study prepares the ground for a wider conversation about sustainability and its crucial role in addressing today's complicated environmental issues.

Call to Action for Stakeholders in Sustainability Efforts

In dealing with sustainability, it is very important to have a unified way that brings together different groups for effective initiatives to improve environmental health. Different areas—like government, business, schools, and community groups—need to go beyond traditional barriers and work together, using their individual strengths and resources. This requires everyone to recognize shared duties, where participants not only give money but also take part in making decisions and sharing knowledge.

By encouraging conversation and understanding, these groups can work together to create new plans that focus on caring for the environment and dealing with social and economic gaps. The pressing issue of climate change pushes everyone to look closely at how they operate and promise to be transparent. Through these combined efforts, the involvement of stakeholders can change from just meeting requirements to taking an active role, helping to establish long-lasting sustainable practices for future generations.

Climate Data—The Backbone of Understanding

Climate change is no longer a distant theory debated in scientific journals. It's a lived reality, shaping weather patterns, ecosystems, and economies across the globe. But how do we know this? How can we track these shifts and predict what might happen next? The answer lies in data. Climate data is the foundation of our understanding, the compass guiding us through an uncertain future. Without it, we'd be stumbling in the dark, trying to address a problem we can't fully see.

The Foundations of Climate Data

Reliable and available data serves as the foundation of climate science. It provides us with the means to comprehend how the Earth is transforming and the reasons behind it. Consider data as the components of a massive puzzle. Every temperature observation, precipitation figure, or satellite photograph links together to create a larger image. Lacking sufficient pieces, we only observe fragments. Our comprehension would be lacking, and our capacity to forecast upcoming trends would be inconsistent. So, where did everything begin? And how did we transition from basic measurements to the sophisticated climate models that we depend on today?

In the early days, climate studies were based on manual observations. Scientists would venture into the field equipped with simple tools to document temperatures, precipitation, and various weather trends. A thermometer provides one measurement at a particular moment. A rain gauge gathered water throughout the day. Barometers monitored variations in air pressure, suggesting changes in the weather. Observations were frequently inscribed by hand in journals, leading to the possibility of them being lost, harmed, or containing mistakes. However, despite their limitations, these initial endeavors represented the first strides toward comprehending our climate.

Imagine standing in a field a hundred years ago, jotting down the day's temperature in a diary. At that moment, it could have seemed insignificant. Only a single digit. Just a moment. However, over the years and decades, those dispersed figures formed the basis for contemporary climate science. Today, we reflect on those handwritten documents with great admiration. They set benchmarks that continue to be utilized for comparison. That information allows us to understand how temperatures, precipitation, and various factors have evolved over time.

The real transformation began with technology. As tools evolved, our ability to observe the planet expanded. Satellites now orbit the Earth, capturing images and measurements we could never have imagined a century ago. Instruments like spectroradiometers detect tiny shifts in surface temperatures and plant health. Remote sensors collect data on rainfall, humidity, and wind. Ocean buoys float on vast seas, measuring currents, water salinity, and sea surface temperatures. Together, they paint a global picture that shows where and how changes are happening.

One of the most groundbreaking developments came with advanced computing. It isn't just about gathering data anymore;

it's about making sense of it. Massive volumes of raw information can now be processed into meaningful insights. Climate models take this data and create simulations of possible futures. They show how a change in one system — like rising ocean temperatures — might trigger shifts elsewhere, such as stronger hurricanes or droughts. These models help us prepare for challenges and build strategies to adapt.

But even with all this progress, gaps remain. Data collection is uneven. Some regions, like major cities or developed countries, have dense networks of sensors and instruments. Other areas, particularly developing nations, struggle to keep up. Deep oceans, polar zones, and remote forests remain hard to monitor regularly. These regions hold important clues about the climate but are often overlooked because of limited resources.

Take the Arctic, for example. It is warming twice as fast as the rest of the planet, but there are fewer permanent monitoring stations there. That's a problem. If we can't consistently measure what's happening in this critical area, we miss key signals about global warming. Scientists are finding creative ways to bridge these gaps. Autonomous drones, deep-sea sensors, and remote buoys are being deployed to explore these hard-to-reach areas. However, solutions like these require funding, collaboration, and time.

There is also the challenge of sharing data. Advanced nations have more satellites and monitoring tools, but not all of this information is freely available. Some countries lack the technological infrastructure to access and use it. A farmer in a drought-stricken region of Africa, for instance, could greatly benefit from weather predictions. Yet without access to reliable rainfall data, they might plant their crops at the wrong time and lose everything. Bridging this divide requires cooperation. Initiatives like the Copernicus program are steps in the right direction. By offering open-access satellite data, they ensure that vital information is available to everyone, not just the privileged few.

At its core, climate data does more than inform scientists. It tells stories. An upward curve on a temperature graph isn't just a line. It's the heatwave that forced families to evacuate their homes. A satellite image of a shrinking glacier isn't just a photo. It's a warning that sea levels are rising, threatening entire communities. Climate data brings these stories to life. It turns abstract trends into real, human experiences that people can see and feel. This power to connect science with everyday life is what makes it so impactful.

However, collecting data is only part of the equation. How we use and interpret it is equally important. The tools we have are powerful, but they can't make decisions for us. A flood prediction might give a community time to evacuate. But will the warning be taken seriously? Will local leaders act in time? These are human choices. The data can guide us, but we have to be willing to listen.

So what happens if we ignore it? What if we decide the data doesn't matter? The consequences are already visible. Extreme weather events, rising sea levels, and shifts in ecosystems are all signals that we can't afford to dismiss. Climate data gives us a chance to act before these changes become irreversible. It alerts us while there is still time to do something about it.

In the end, data is our compass. It shows us where we are, where we're headed, and what we can do to change course. The question is whether we will pay attention. The tools are in our hands, and the answers are within reach. What we choose to do with them is up to us.

Technologies and Tools

Contemporary climate science relies on an advanced system of technologies and instruments to collect, process, and evaluate data. These instruments offer the in-depth information required to comprehend the planet's transformations and predict potential future developments. From satellites circling the Earth to ground stations situated in isolated regions, every component of technology is essential in enhancing our understanding.

Remote Sensing

Satellites have entirely transformed our methods of observing the Earth. Prior to their creation, researchers depended on scattered ground measurements, which provided restricted data. Currently, satellites gather extensive data regarding the Earth's surface, atmosphere, and oceans, providing a global viewpoint that was previously inconceivable.

Consider, for instance, NASA's Terra and Aqua satellites. These satellites provide extremely detailed pictures of the Earth's surface, monitoring alterations in temperature, plant life, and cloud formation. This information enables researchers to identify trends such as desert growth or deforestation that could otherwise remain hidden. In the same way, the Sentinel series from the European Space Agency provides high-resolution imagery that tracks ice sheet depletion, changes in land use, and increases in sea levels. These instruments assist us in monitoring worldwide trends swiftly and with remarkable precision.

However, remote sensing encompasses much more than mere images. Instruments such as the Advanced Very High Resolution Radiometer (AVHRR) and the Moderate Resolution Imaging Spectroradiometer (MODIS) collect information on temperature, radiation, and various other essential elements. This information is crucial for comprehending the behavior of the climate at both global and regional levels.

LIDAR technology is one of the most remarkable instruments in remote sensing. LIDAR employs laser pulses to gauge surface elevation with remarkable accuracy. This has been crucial in polar areas, as it monitors the reduction of ice sheets. For instance, LIDAR data from 2002 to 2020 indicated that the rate of ice loss in Greenland was increasing more rapidly than earlier estimates suggested. These measurements act as undeniable proof of how climate change is altering the Earth.

Rainfall tracking is another potent use of satellite technology. The Global Precipitation Measurement (GPM) mission gathers real-time information on global rainfall trends. This information is essential for areas such as sub-Saharan Africa, where agriculture relies on comprehending rainfall fluctuations. The GPM mission enables communities to brace for weather extremes and safeguard their crops and livelihoods by forecasting floods or droughts.

Ground Stations

While satellites offer a perspective from above, ground stations deliver the local, on-site information required to validate satellite findings. These stations consist of weather stations, ocean buoys, and terrestrial sensors that gather accurate data on temperature, precipitation, wind velocity, and humidity. Consider them as links that connect extensive global insights to actual, localized circumstances.

Ocean buoys, for example, are crucial for comprehending marine ecosystems. These floating instruments measure the temperature of the sea surface, salinity, and currents—information that provides insight into the condition of our oceans. The Argo initiative is a prominent illustration. It employs a system of self-operating floating sensors that descend to significant depths, gather essential data, and subsequently resurface to relay their results. These sensors have played a crucial role in verifying that ocean heat content reached unprecedented levels in 2022, providing compelling evidence of persistent climate change.

Radar systems constitute another essential component of ground-based monitoring. Doppler radar, commonly utilized in meteorology, monitors storms, tornadoes, and hurricanes with impressive precision. Scientists can provide earlier and more accurate warnings for extreme weather events by integrating radar data with satellite observations. This protects lives by allowing communities additional time to get ready and evacuate when needed.

Known Good Satellites

Some satellite missions stand out because of their reliability and long-term contributions to climate science. These "known good" satellites have become benchmarks for accurate and consistent data collection.

The Geostationary Operational Environmental Satellites (GOES) serve as an excellent illustration. These satellites constantly observe weather patterns, aiding scientists in monitoring hurricanes, storms, and other severe phenomena. Throughout Hurricane Ida in 2021, GOES satellite images were essential for forecasting the storm's trajectory and coordinating extensive evacuations.

Copernicus, the primary Earth observation initiative of the European Union, is also a front-runner. The Sentinel-5P satellite is dedicated to observing air quality. Throughout the COVID-19 lockdowns, Sentinel-5P demonstrated significant declines in nitrogen dioxide concentrations, providing a vivid illustration of the impact of human actions on air quality. This information not only guides climate policies but also indicates how rapidly certain environmental changes can take place.

Japan's Himawari satellite series specializes in high-frequency imaging. It provides almost immediate updates on weather patterns in the Asia-Pacific area. During Typhoon Haiyan, Himawari's swift data delivery provided officials with essential information, enabling them to respond rapidly and save numerous lives. Its high-definition images guarantee that even minor shifts in weather patterns are identified and examined.

These satellites are not merely standalone instruments; they are components of a worldwide network. Through collaboration, they guarantee that data gathering stays uniform and dependable. If a satellite fails, another can frequently take its place, decreasing the chance of losing vital information. This repetition fosters confidence in the data and enables tackling the intricate issues of climate change with increased assurance.

Applications of Climate Data

Climate data is more than just numbers on a screen. It's the key to understanding trends, preparing for natural disasters, and creating informed policies that shape the future. The insights it provides drive decisions that protect lives, economies, and ecosystems.

Tracking Climate Trends

Information allows us to see the transformations occurring in the world. Temperature data, rainfall patterns, and sea-level statistics indicate changes that affect all areas of the globe. For instance, extensive temperature records from the National Oceanic and Atmospheric Administration (NOAA) indicate a continuous increase in global temperatures over the years. This increase in temperature impacts weather patterns, causes glacier melt, and influences ocean ecosystems. Satellite-based sea-level monitoring, utilizing systems such as Copernicus and TOPEX/ Poseidon, informs us about the risks posed by rising waters to coastal regions. Understanding these trends is crucial. It enables researchers to recognize trends and forecast what is to come.

However, what do these trends signify for communities? A tiny island country encountering higher sea levels can prepare for relocations or enhance its defenses using this information. Farmers facing erratic rainfall can modify their crops and irrigation methods. Monitoring these changes is not solely about comprehending the Earth—it's also about safeguarding the individuals who reside on it.

Disaster Prediction and Management

Climate information is essential when the forces of nature become harmful. Hurricanes, floods, and wildfires are increasingly common and severe. Without information, communities would find themselves unprepared for such disasters. Satellites such as GOES and Himawari offer alerts in advance by tracking the formation and movement of storms. Radar systems on the ground monitor severe weather occurrences, providing communities with time to get ready.

Reflect on Hurricane Katrina from 2005. Sophisticated weather data forecasted its trajectory and strength, leading to evacuations that preserved numerous lives. In areas susceptible to wildfires, such as California, immediate data from thermal imaging satellites detects hot spots prior to the spread of fires. Models for flood prediction, which rely on rainfall and river flow information, enable countries such as Bangladesh to reduce losses during monsoon periods.

Disaster information goes beyond just forecasts. It also directs efforts for recovery. Understanding the locations of the most damage assists responders in distributing resources efficiently. This mix of anticipation and after-action makes data an essential asset for managing disasters.

Data in Policymaking and Climate Negotiations

Strategies aimed at addressing climate change require strong foundations. Leaders depend on climate information to make well-informed choices. The Paris Agreement, a significant milestone in worldwide climate discussions, was influenced by years of studies and information from organizations such as the

Intergovernmental Panel on Climate Change (IPCC). Their findings, derived from data gathered in hundreds of studies, indicated an immediate necessity to curb warming.

Data-driven policies are not only global; they are also local. A city struggling with air quality could utilize satellite images to pinpoint pollution hotspots and implement tougher emissions regulations. A nation facing regular droughts may adopt water conservation technologies upon analyzing precipitation patterns.

In climate discussions, information bolsters cases. Nations showing proof of increasing sea levels or land degradation can build a more compelling argument for global assistance. Information also holds entities responsible. Consistent reports guarantee that countries adhere to their obligations, fostering trust and transparency in contracts.

The Bigger Picture

Climate data does not function in isolation. It links science, communities, and governments. It assists in monitoring change, equips us for difficulties, and guarantees our policies are based on evidence. This network of information provides humanity with a viable opportunity to face the challenges of a changing planet.

Should we cease monitoring? Or ought we to utilize all available information to make wiser choices? The decision is obvious. Climate information serves as our compass. Let's ensure that we pay attention.

Challenges in Climate Data

The power of climate data lies in its ability to inform and inspire action, but it is not without challenges. These obstacles, if left unaddressed, can hinder progress and limit the potential of data to drive meaningful change.

Accessibility and Equity in Data Sharing

Not all regions have equal access to climate data. Wealthier countries often lead in satellite launches, ground station networks, and data analytics. Meanwhile, many developing nations struggle to access this wealth of information. Without proper data, these regions face greater challenges in preparing for climate impacts.

Imagine a country dealing with unpredictable monsoons but lacking rainfall data. Farmers can't make informed decisions, and governments can't allocate resources effectively. Bridging this gap requires global cooperation. Initiatives like the World Meteorological Organization's Climate Services aim to ensure data reaches those who need it most. The Copernicus program, which shares open-access satellite data, is another step toward closing this gap. Still, the digital divide and lack of resources in many regions show how far we need to go.

In addition, inequitable data sharing often leaves vulnerable communities at greater risk. For example, regions in sub-Saharan Africa depend heavily on agriculture but lack consistent weather monitoring systems. Real-time data from tools like rain gauges or satellite imaging could save crops during droughts or floods. Yet, the infrastructure to collect and use this information often lags behind.

Dealing with Gaps and Inconsistencies

Despite having advanced technologies, gaps in data still remain. Isolated areas, vast seas, and Arctic regions frequently receive insufficient monitoring. The Arctic, for instance, is heating up at double the global pace, but the area does not have enough permanent ground stations to monitor this transformation. Scientists have adopted cutting-edge techniques such as unmanned underwater vehicles and remotely operated buoys to address some of these gaps. Initiatives like the IceBridge Mission utilize sensors attached to aircraft to examine regions that are challenging for satellites, though such endeavors necessitate substantial funding and collaboration.

Variable measurements also hinder analyses. Data gathered in the past with less accurate tools might not completely match current datasets. This problem is particularly clear when merging historical climate data with satellite-based information. Tools for machine learning are increasingly utilized to close these gaps, aiding researchers in forecasting absent values and synchronizing diverse datasets. For instance, scientists employed AI to assess ocean heat content in periods when direct measurements were limited. Although there have been improvements, gaps and inconsistencies continue to pose challenges, highlighting the necessity for cooperative efforts among nations, academic institutions, and the private sector.

Ethical Considerations in Data Usage

Data has the power to drive change, yet it can also be exploited. There is a danger of data being selectively shown to reinforce biased stories. For example, selectively using statistics to minimize the significance of climate change can deceive the public and hinder advancement. It is crucial to guarantee transparency and accountability in the way data is presented. Governments and organizations must implement open-data policies that ensure climate information is accessible to all rather than only specific groups.

Another issue is privacy. As monitoring technologies evolve, concerns emerge regarding the equilibrium between collecting data and honoring individual or community rights. For instance, drones employed to observe deforestation might unintentionally record personal or sensitive details regarding nearby communities. Ethical frameworks need to develop in tandem with technology to tackle these issues. Entities such as the United Nations' Global Pulse are investigating methods to utilize big data ethically, prioritizing ethical concerns throughout the process.

The Way Forward

Openly sharing climate data can transform the global reaction to climate change. Open data initiatives guarantee that information is accessible and not confined behind paywalls or restricted to select groups. By providing access to data for all, we enable scientists, policymakers, and communities to make well-informed choices.

The accessibility of climate data enables local solutions to develop. For instance, Kenyan farmers utilize rainfall information from public platforms to strategize planting and watering. In the Philippines, community organizations also utilize free mobile apps that combine satellite and weather station data to access early warning systems for typhoons. These instruments convert raw data into practical insights that protect lives and incomes.

Cooperation is central to this initiative. Initiatives like the Climate Data Initiative, spearheaded by agencies like NASA and NOAA, provide extensive datasets to researchers worldwide. This method dismantles barriers, promoting creativity. Envision scientists in one nation creating a flood forecasting model while another group enhances it for drought control. When information is exchanged, advancement turns into a joint endeavor.

The advantages of open data go beyond just research. It fosters responsibility in climate discussions. Nations can confirm one another's emissions statements, guaranteeing that pledges are fulfilled. This openness builds trust, which is crucial for international collaboration.

Obstacles persist, yet answers are attainable. Investing in infrastructure can bridge the digital gap, whereas strategies that encourage data sharing can remove obstacles. Ethical principles can guarantee responsible data usage and safeguarding privacy while fostering innovation.

The way ahead is based on a straightforward principle: information ought to benefit all. By adopting open data, we foster a reality where information inspires action. It's a realm where answers arise not just in boardrooms but also in communities, labs, and classrooms equally. The issue now isn't whether we can provide access to climate data but rather whether we decide to do so. The response will mold our shared future.

Climate Intelligence— Transforming Data into Action

As we move forward, it's not enough to simply gather data. The subsequent step is to utilize that data in ways that promote significant action. Climate intelligence combines science and technology, converting raw data into effective tools for informed decision-making. It's focused on transforming climate information into practical insights that can save lives, preserve resources, and create a sustainable future. In a world progressively influenced by environmental issues, utilizing data wisely can determine the line between readiness and catastrophe.

What Is Climate Intelligence?

Climate intelligence utilizes the extensive climate data gathered from satellites, ground stations, and various sources, employing sophisticated tools for its analysis. Artificial intelligence, machine learning, and predictive analytics are central to this change. These technologies analyze data more rapidly than any human could, identifying patterns and trends that may otherwise remain undetected. The outcome is a more distinct understanding of current events and potential future developments.

Envision monitoring precipitation for years to foresee droughts or studying wind trends to enhance renewable energy efficiency.

AI and machine learning are proficient at recognizing these intricate relationships. In doing this, they enable governments, enterprises, and communities to get ready for challenges and capitalize on opportunities. When these tools are utilized, they offer not only solutions but also clarity, assisting individuals in making decisions with increased confidence.

Think about the advancements made in weather forecasting. Conventional models depended on historical averages and fundamental physics. Currently, predictive algorithms utilize real-time information, assisting meteorologists in delivering more precise forecasts. These developments do more than forecast the weather. They also forecast its effect. For example, predictive analytics can assess the financial impact of a storm prior to its arrival, helping with evacuations and resource distribution. Emergency planners are able to respond more quickly, preserving both lives and resources.

Climate intelligence surpasses mere forecasting. It's about discovering more intelligent methods to address issues. AI assists in enhancing urban planning, guaranteeing that cities are constructed with resilience as a priority. Machine learning algorithms enhance energy efficiency by forecasting demand and minimizing waste. Predictive analytics provide farmers insights on optimal planting and harvesting times, enhancing yields while preserving water. These instruments are transforming the ways industries and communities tackle sustainability.

The farming industry is currently reaping advantages. AI-powered systems assist farmers in observing soil conditions, monitoring weather variations, and regulating irrigation. In India, for example, small farmers utilize mobile applications that leverage predictive analytics to identify the optimal times for planting seeds. This method enhances crop production while also minimizing water and fertilizer consumption, resulting in more sustainable agricultural practices.

City regions are likewise adopting climate awareness. Cities such as Singapore utilize AI to oversee energy usage instantaneously. Intelligent grids forecast energy needs and modify distribution, minimizing waste and decreasing emissions. In regions prone to disasters, urban areas employ predictive analytics to identify possible flood zones or wildfire routes, allowing residents the opportunity to evacuate and get ready. These instances demonstrate that climate intelligence is more than just a theoretical idea; it actively influences daily decision-making processes.

The real strength of climate intelligence is its capacity to make connections. It involves leveraging technology to convert data into strategies and transforming information into meaningful results. The instruments are already present. The issue is how we decide to utilize them. Will we take decisive action to fully leverage these technologies, or will we allow chances to fade away? The decisions we take today will influence the future of our adaptation to a shifting climate.

Applications in Action

Climate intelligence is not an abstract idea. It is already making a difference across sectors, turning data into meaningful actions with measurable results.

Disaster Preparedness

Climate insight provides communities with a vital edge in facing disasters. Accurate forecasting of storms, floods, and wildfires protects lives and valuables. Satellite data integrated with AI models detect high-risk regions prior to disaster occurrences. This data enables officials to respond, whether by providing advance alerts or evacuating at-risk communities.

For instance, in Bangladesh, AI is utilized in flood forecasting systems to evaluate rainfall and river flow information. These systems offer alerts as early as ten days ahead, allowing individuals to safeguard their homes and relocate to safety. In California, predictive tools assess wildfire risks by examining wind speeds, temperatures, and vegetation types. Firefighters utilize this information to distribute resources to areas in greatest need, stopping smaller blazes from escalating uncontrollably. In the same way, the Australian government utilizes AI-based systems to forecast bushfire threats, combining weather information and satellite images to respond promptly during fire periods.

Disaster readiness involves more than just reaction. It also involves planning for the long term. Hurricane-prone cities can enhance their infrastructure by utilizing predictions of storm impacts. Coastal communities can utilize sea-level rise projections to determine locations for constructing flood barriers. In Japan, climate awareness has influenced the creation of tsunami barriers, helping to better safeguard communities from future calamities. Climate intelligence shifts reactive approaches into proactive ones, lessening the enduring impact of disasters.

Urban Planning

Cities are at the forefront of climate challenges. Rising temperatures, flooding, and extreme weather events make traditional urban planning insufficient. Climate intelligence offers a smarter way forward. By integrating predictive analytics, urban planners can design cities that withstand environmental pressures.

Singapore's Green Plan is a prime example. The city uses AI-powered systems to optimize energy use in buildings, reducing emissions while maintaining comfort. Sensors track the temperature and air quality, providing real-time data to improve living conditions. Predictive models help design flood-resistant infrastructure by simulating rainfall patterns and water flow. In addition, Singapore's AI tools guide green roof installations to reduce urban heat, enhancing energy efficiency.

In the Netherlands, climate intelligence has shaped how cities manage rising sea levels. AI tools analyze storm surges and erosion risks, guiding the construction of resilient dikes and barriers. Rotterdam's water plazas—public spaces that double as flood control systems—were developed using these insights. By combining innovation with data, urban areas can adapt to challenges and enhance quality of life. Elsewhere, cities like Copenhagen are leveraging climate intelligence to redesign public transport routes, reducing emissions and easing congestion during extreme weather.

Agriculture

Farmers face growing uncertainty as climate patterns shift. Climate intelligence helps them adapt by providing detailed forecasts and actionable insights. AI-driven tools monitor soil health, track pests, and predict weather changes, giving farmers the information they need to make better decisions.

In Kenya, mobile platforms powered by climate intelligence send daily weather updates to farmers. These platforms also provide advice on planting schedules and crop choices based on predicted rainfall. The results are clear. Yields improve, and losses from droughts or floods are minimized. Similar systems in India have helped smallholder farmers optimize water use, reducing both costs and environmental impact. For instance, the Maharashtra Water Resources Department has implemented predictive analytics to manage water distribution for crops during dry spells, helping millions of farmers.

Precision agriculture is another area where climate intelligence excels. Sensors and drones collect data on crops, which is then analyzed to optimize irrigation and fertilization. This reduces waste and boosts productivity. In the United States, large-scale farms are using AI models to predict yield outcomes and plan harvests more efficiently. In Brazil, AI-powered tools monitor deforestation impacts on crop production, allowing farmers to adjust practices for better environmental outcomes. Climate intelligence turns farming into a data-driven practice, making it more sustainable and resilient.

Corporate Strategy

Businesses are increasingly turning to climate intelligence to stay competitive in a changing world. For corporations, it's about understanding risks and opportunities. Predictive models help companies assess how climate change will affect supply chains, energy needs, and market demands.

Take the insurance industry. Climate intelligence helps insurers predict the financial impact of disasters, enabling them to set premiums more accurately. For example, AI models that analyze hurricane paths and flood risks allow insurers to calculate potential payouts with greater precision. This protects their bottom line while ensuring policyholders are better prepared. In the UK, climate intelligence tools assess flood risks for property insurance, helping residents in high-risk zones receive tailored coverage.

In the energy sector, companies are using climate intelligence to optimize operations. Wind farms analyze weather patterns to maximize energy generation. Solar power providers use predictive tools to estimate output based on cloud cover and temperature. These insights improve efficiency and reduce costs. In Europe, utilities are leveraging AI to predict energy demands during heatwaves, preventing blackouts and maintaining grid stability.

Retailers are also utilizing climate intelligence. Through examining temperature and seasonal trends, they modify inventory to align with consumer demand. For example, a clothing retailer could maintain a smaller inventory of winter coats in areas with warmer winter climates. Major food retailers utilize climate models to forecast crop deficits, allowing them to source items more efficiently and minimize supply chain interruptions.

Climate intelligence is aiding companies in progressing past short-term perspectives. It enables them to prepare for a future influenced by environmental change. Whether it's mitigating risks or uncovering new prospects, firms that adopt climate intelligence are more equipped to succeed. By employing the appropriate tools and strategies, sectors can transform climate-related challenges into opportunities for innovation and expansion.

Leading Climate Intelligence Platforms

Climate intelligence is not just theoretical; it is already making waves through innovative platforms and collaborative efforts. Several organizations are taking charge of transforming climate data into actionable tools.

Ambee, my climate intelligence startup, incorporates real-time information on air quality, temperature, and humidity into its system. This information is utilized to fuel applications across various sectors, including healthcare and logistics. For instance, logistics firms utilize Ambee's insights to enhance delivery routes, reducing exposure to bad air quality and extreme temperatures for both employees and products. Cities utilize Ambee's tools to create improved pollution control strategies.

Another significant instance is IBM's Environmental Intelligence Suite. This platform integrates AI, meteorological information, and spatial analytics to assist companies in getting ready for environmental hazards. Retailers utilize it to predict supply chain interruptions during extreme weather incidents, whereas energy firms depend on it to optimize resource management. A case study illustrates how the suite aided a solar energy provider in forecasting cloud cover, more precisely, increasing energy production by over 15% within a year.

Climate TRACE, an international alliance, monitors greenhouse gas emissions almost in real time by utilizing satellites, artificial intelligence, and information from numerous sensors. By pinpointing particular emission sources, like factories or transportation centers, Climate TRACE delivers the clarity necessary to hold polluters responsible. Governments have started utilizing this data to develop more focused climate policies, making certain that emission reductions are not only pledged but also quantifiable.

These platforms are merely the start. Worldwide, startups and governments are incorporating climate intelligence tools into their functions. In Africa, farmers receive customized weather updates and crop guidance through mobile apps. In Europe, local governments employ AI-driven flood mapping to create safer communities. These tools signify a transition from reactive responses to proactive approaches, demonstrating that climate intelligence goes beyond being merely a buzzword, it's a catalyst for change.

Technical Limitations

Advanced climate intelligence relies on cutting-edge technology, but these tools often require significant infrastructure and expertise. Many regions lack access to high-speed internet or the hardware necessary to run complex models. Small organizations and underfunded governments may struggle to afford or maintain the tools needed to harness climate intelligence effectively.

Even in well-equipped areas, gaps in climate data can reduce the accuracy of models. Remote sensing satellites can't always capture fine details, especially in remote areas like deep oceans or dense forests. Ground stations may be sparse, leading to inconsistencies in the datasets used for analysis. Without reliable inputs, even the most sophisticated AI models can fall short.

Data Biases

Biases in climate data can lead to skewed results, which may harm decision-making processes. Some datasets overrepresent developed regions where monitoring infrastructure is dense, while underrepresenting poorer or rural areas. This creates blind spots that leave vulnerable communities without the insights they need to prepare for climate impacts.

Another concern is algorithmic bias. AI systems learn from existing datasets, which means that historical biases in the data can carry forward into predictions. If past decisions prioritized urban centers over rural ones, future models might continue to do so, compounding inequities instead of addressing them.

The Digital Divide

Access to climate information is not evenly provided. Numerous developing countries and rural regions do not possess the technological infrastructure required to take advantage of these tools. In the absence of dependable internet connectivity, sophisticated equipment, or skilled staff, vast areas are excluded from the climate intelligence network. This gap worsens current disparities, leaving the most at-risk groups ill-equipped to face climate challenges.

For instance, smallholder farmers in sub-Saharan Africa might not have access to weather forecasting tools that could assist them in planning their crops. Coastal communities in Southeast Asia may lack access to early flood warnings because of inadequate connectivity. Closing this divide necessitates investment in infrastructure, education, and collaborations between governments and private entities. Initiatives such as the Global Partnership for Sustainable Development Data are advancing, yet the speed must quicken to guarantee that everyone is included.

The Need for Multidisciplinary Collaboration

Solving climate challenges with intelligence tools demands collaboration across disciplines. Scientists, technologists, policymakers, and local communities all bring unique perspectives and skills. Without this cooperation, solutions risk being siloed, incomplete, or impractical.

The partnership between climate scientists and AI developers, for example, has already enhanced predictive models for severe weather. Collaborations between governments and technology firms have facilitated improved data-sharing platforms, like NASA's Earth Science Data Systems. Engaging local communities guarantees that solutions are pertinent and culturally suitable. For instance, projects led by communities in India have combined traditional agricultural knowledge with AI-assisted irrigation technology, resulting in better results for farmers.

Establishing these partnerships is not always straightforward. Various fields frequently possess distinct priorities and terminologies, which complicates communication. However, the advantages of collaborating surpass the challenges. Establishing frameworks that promote conversation, financing collaborative initiatives, and nurturing trust among stakeholders can dismantle these obstacles. Climate emergency is a worldwide problem, and addressing it requires everyone's involvement.

These obstacles might appear overwhelming, yet they are not impossible to overcome. By tackling technical constraints, removing data biases, closing the digital gap, and promoting cooperation, we can realize the complete potential of climate intelligence. The issue isn't if we possess the tools but rather if we can guarantee that those tools are accessible to all and utilized to their maximum effectiveness.

The Future of Climate Intelligence

The development of climate intelligence will be driven by cooperation, innovation, and inclusiveness. Combining citizen science, local insights, and cutting-edge AI will be crucial in enhancing the effectiveness and accessibility of climate intelligence.

Citizen science is currently having an influence. Volunteers globally provide information on air quality, water levels, and wildlife observations, bridging deficiencies in official documentation. Websites such as iNaturalist and Global Forest Watch depend on user submissions to develop comprehensive environmental maps. These community-led initiatives gather localized information that is frequently overlooked by extensive monitoring systems, contributing to a better understanding of environmental shifts.

Citizen scientists in the U.S., for instance, have assisted in monitoring butterfly migrations, offering essential information about the impacts of climate change on insect communities. Comparable initiatives in Europe include tracking air quality using inexpensive sensors given to volunteers and generating hyper-local pollution maps that enhance official data sets. These efforts demonstrate how everyday individuals can impact worldwide decision-making.

Knowledge from the community is just as crucial. Indigenous communities, for instance, have a profound knowledge of their ecosystems. In the Arctic, Indigenous communities have collaborated with scientists to monitor sea ice trends, merging traditional insights with satellite information for better accuracy. In the Amazon, Indigenous monitors utilize GPS technology to chart deforestation, offering ground-based verification that enhances satellite images. Fishers in Southeast Asia have partnered with scientists to understand how shifts in ocean temperatures affect fish migration, developing adaptive methods for sustainable fishing.

Advanced AI will link all these inputs together. Through analyzing large volumes of varied data, AI systems can reveal patterns that were once hidden. For instance, AI systems might combine citizen reports of floods with rainfall information and satellite imagery to forecast upcoming high-risk locations. Authorities and institutions can subsequently function more swiftly and with enhanced accuracy. Algorithms for machine learning are currently employed to assess wildfire threats in Australia by integrating real-time weather information, historical fire trends, and vegetation data to improve response tactics.

Cooperation between international and local participants will be crucial. Governments can establish systems that promote citizen engagement, while technology firms can develop tools that are user-friendly for those without expertise. When cutting-edge AI collaborates with community initiatives, climate intelligence will achieve its maximum potential. This integration will guarantee that every voice is acknowledged and no data point is missed. For example, the OpenStreetMap project, led by volunteers and backed by various organizations, has played a crucial role in mapping at-risk areas for climate resilience strategies.

The future of climate awareness involves more than just technology. It focuses on establishing connections—among individuals and systems, among worldwide initiatives and local efforts. United, we can build a more intelligent and sustainable future. This future requires a collective dedication from every part of society. No matter if it's a farmer providing data via a mobile app or a tech developer improving AI models, each input is important. It is this joint endeavor that will determine our response to the climate challenges that lie ahead.

Closing Note

As we arrive at the end of this book, we stand at the threshold of a new era—one in which the realities of climate change are no longer distant warnings but lived experiences unfolding around the globe. Each chapter you've explored has touched on critical facets of our planetary crisis, offering insights into the scientific, social, economic, and ethical complexities that shape our response. From the concept of tipping points and the perilous brink of cascade to the promise of new technologies and the power of global cooperation, these pages have guided you through the tumultuous landscape we inhabit. Now, with a deeper understanding of both the challenges and opportunities, we can synthesize this knowledge into a cohesive vision of hope and action. This concluding chapter aims to consolidate the themes we've traversed, leaving you with a sense of fulfillment, empowerment, and readiness to step forward as an agent of change.

Weaving the Threads of Urgency and Action

From the outset, the **introduction impressed upon us the reason** climate change matters. It is not merely an abstract scientific concept but a tangible force with immediate effects on human well-being, economic stability, and environmental integrity. Every heatwave, every wildfire, every superstorm that has disrupted lives in recent years testifies to the urgency of the crisis. It's clear that climate change is not a future problem; it's happening now, affecting our food supplies, public health, and geopolitical landscapes in real-time.

More than simply diagnosing problems, this book sets out to equip you with actionable knowledge. And throughout our journey, we've seen how individuals, communities, policymakers, and businesses can each play pivotal roles. The chapters were structured to provide a building-block approach: first establishing the foundational science, then examining human influences, discussing global policy frameworks, showcasing technological and social solutions, and ultimately envisioning a future shaped by collective willpower. In reading these pages, you have immersed yourself in the interconnected nature of climate challenges, culminating in a mosaic of perspectives that demonstrate the multifaceted nature of climate solutions.

Revisiting the Brink of Cascade

Chapter 1: The Brink of Cascade underscored how climate tipping points operate like hidden tripwires, where a small additional push can unleash outsized, irreversible impacts. We explored historical climate events such as the Younger Dryas, a period when a sudden influx of freshwater into the North Atlantic disrupted ocean currents and brought an abrupt shift in global temperatures. These historical precedents are valuable cautionary tales; they remind us that once we pass certain points, returning to stability becomes exponentially harder.

This chapter also introduced the notion of the "Last Degree," emphasizing how critical the final increments of temperature rise can be. Whether it's melting permafrost unleashing vast stores of methane or the destabilization of polar ice sheets that accelerate sea-level rise, these near-invisible processes at the planet's extremes can reverberate throughout the world. The ripple effects—such as changes in thermohaline circulation—highlight how climate systems do not exist in silos. Melting in the Arctic can spell chaotic weather patterns in the tropics and temperate zones, illustrating the intricate web of connections that bind local and global climates.

Reflecting on this, the lesson is clear: If we are to stave off catastrophic changes, we must keep global warming well below those critical degrees. Each fraction of a degree matters for the ecosystems and communities most vulnerable to disruptions. In all of this, the underlying message is one of prevention: once the cascade begins, halting it becomes immensely more difficult and costly.

The Human Factor: Our Greatest Challenge, Our Greatest Hope

Chapter 2: The Human Factor brought us face-to-face with our own responsibility in accelerating climate change. Industrialization, agriculture, and urban sprawl have all contributed to atmospheric carbon buildup. Yet, pointing fingers at "humans" as a monolith misses the nuances of global inequalities. The chapter reminded us of our moral and ethical responsibilities, emphasizing that we are not merely passive consumers but active stewards—or, alternatively, disruptors—of Earth's life-support systems.

We dissected the role of carbon emissions across sectors, from factories and power plants to cars and airplanes, all while exploring the intricate relationship between modern lifestyles and climate impact. Psychological factors also emerged as potent forces shaping public and political will. Denial, fear, and helplessness can paralyze efforts to act, while hope and empowerment can kindle transformative movements. The chapter also contrasted traditional agriculture's heavy resource footprint with sustainable alternatives—permaculture, organic farming, and vertical agriculture—that harness innovation to reduce water usage, chemical inputs, and biodiversity loss.

Ultimately, the human factor is twofold: we are the source of much of the climate problem, but we also hold the solution. By making mindful choices, advocating for policy shifts, and championing climate considerations in all areas of life, we can pivot from being the problem to becoming the driving force of positive change.

Policy, Diplomacy, and Global Cooperation: The Glue That Binds Action

Chapter 3: Policy, Diplomacy, and Global Cooperation zoomed out to the broader stage of international negotiations and treaties. It highlighted how global initiatives—like the **Paris Accord**—came into being after decades of incremental progress, setbacks, and renewed resolve. We traced climate policy evolution, acknowledging that while the Paris Accord was a watershed moment in global diplomacy, its success hinges on the collective ambition of nationally determined contributions (NDCs) and the accountability mechanisms to measure and enforce them.

We also examined the challenges that stand in the way of unified global action. Geopolitical tensions, economic disparities, and uneven historical responsibilities for emissions can fracture global unity. Developing nations often argue they have the right to develop their economies and lift their citizens out of poverty, while industrialized nations grapple with the legacy of having contributed most of the carbon already warming the planet. The concept of climate justice thus resonates powerfully, highlighting that any equitable climate solution must account for these historical and ongoing imbalances.

In the face of these challenges, collaborative platforms like the Conference of the Parties (COP) meetings serve as critical venues where countries, businesses, civil society organizations, and activists converge. The emergence of new global initiatives such as carbon trading systems, Green New Deal proposals in various regions, and local government leadership illustrate that solutions are not the monopoly of any single actor. Rather, it is a patchwork of multi-level efforts, each reinforcing the other, that can propel us forward.

Voices of Change: Grassroots to Global

Chapter 4: Voices of Change spotlighted the growing chorus of individuals and organizations pushing for immediate, effective climate action. From indigenous communities safeguarding their ancestral lands in the Amazon to youth activists using social media to mobilize millions, this chapter illustrated that policy pronouncements alone are not enough. Real transformation is fueled by a groundswell of civic engagement, moral clarity, and collective determination.

We saw how local grassroots movements often spearhead broader societal shifts. The synergy between top-down policies and bottom-up activism can spark meaningful change, whether through rewilding projects that restore entire ecosystems or community-led initiatives that reimagine urban living. Funding, too, is flowing toward sustainable solutions as more investors recognize the risks of a warming world. **The Investment Shift** we covered underscores that when major investment funds, banks, and venture capitalists pivot toward green technologies and social impact, the entire market can realign with sustainable principles.

The lesson here is that you don't need to be a head of state or a billionaire CEO to ignite change. Each person's voice has value, especially when it's part of a larger collective call. Every local initiative, every social media campaign, and every community project adds another layer of momentum that pushes political will in the right direction.

Turning the Tide: Imagining Our Shared Tomorrow

Chapter 5: Turning the Tide guided us through possible future scenarios. The chapter painted two divergent paths: **the world we might inherit** if current policies and complacencies persist and **the world we could build** if innovation and political resolve coalesce in time. Drawing on IPCC reports and emerging scientific research, the message was clear: humanity stands at a crossroads. The decisions made in the next decade will significantly shape the trajectory of our planet for centuries.

We explored how economics and sustainability are increasingly intertwined, with circular economies and resilient infrastructures taking root across the globe. In the Netherlands, urban areas have adapted to constant flooding threats by designing waterborne neighborhoods and multi-purpose dikes that double as recreational spaces. Elsewhere, wildfire-prone regions invest in fire-resistant building materials, proactively adjusting to a "new normal" of more frequent and intense fires. These stories of adaptation underscore that, while mitigation of emissions is critical, preparing for the climate disruptions already in motion is equally essential.

At the same time, the transformative power of culture and politics offers hope. Countries that were once reluctant to adopt ambitious climate targets are increasingly realizing that sustainability can be a potent economic driver. Renewable energy sectors employ large workforces, and circular economies can spur local manufacturing and entrepreneurship. Thus, transitioning to a low-carbon future isn't a burden—it can be a defining economic opportunity that re-invigorates communities and fosters innovation.

Mitigation Strategies and Solutions: Practical Paths Forward

Chapter 6: Mitigation Strategies and Solutions delved deeper into specific ways we can address climate change head-on. One of the standout insights was the role of emotional and mental health—particularly **eco-anxiety**—in either stoking or stifling climate action. Recognizing that feelings of despair or overwhelm can paralyze people, some organizations and communities have turned to group therapy, nature retreats, and resilience workshops to channel that anxiety into action. This might sound peripheral to greenhouse gas inventories or policy papers, but mental health can be the linchpin that either enables sustained activism or leads to burnout.

Corporate responsibility also took center stage, with examples of companies incorporating climate considerations into their wellness programs and supply chains. While skepticism about "greenwashing" remains warranted, there is no question that a growing segment of the business world sees sustainability as a strategic imperative rather than a mere marketing buzzword.

On the ground, we encountered numerous examples of resilience building: green roofs for reducing urban heat, water management systems for tackling flooding, and reforestation for carbon capture and biodiversity conservation. Indigenous-led stewardship projects highlighted how traditional ecological knowledge can guide land management in ways that Western practices have often overlooked. From safeguarding wetlands to supporting pollinators, these nature-based solutions reinforce that restoring ecosystems is not a quaint ideal but a powerful strategy for mitigating climate impact, preserving biodiversity, and supporting human livelihoods.

The Role of Technology and Innovation: Bridging Empathy and Systems Change

Chapter 7: The Role of Technology and Innovation illuminated how powerful tools—from virtual reality experiences that foster empathy to advanced renewable energy systems—can drive a green transformation. We learned of projects like VR simulations that let participants feel the stark consequences of climate change firsthand, whether that's standing in a deforested tropical forest or facing down a melting glacier. By making these scenarios visceral, technology can break through the psychological barriers that often keep people complacent.

Simultaneously, we surveyed how urban centers are undergoing a **Green Metamorphosis**, with smart grids, green buildings, and eco-friendly public transportation becoming more commonplace. Cities like Copenhagen, Singapore, and Curitiba have proven that concentrated populations can be an advantage: greater efficiency in energy use, easier access to mass transit, and the potential for large-scale communal projects. We also touched on **Innovative Water Management Technologies**, an increasingly urgent field as drought and water scarcity intensify across continents. Smart irrigation, water recycling, and atmospheric water capture are among the solutions helping regions adapt while also reducing resource strain.

In the realm of **Energy Revolution**, we saw how solar panels, wind turbines, and new horizons such as tidal energy are poised to redefine the global energy mix. The pace of technological improvement, coupled with falling costs, suggests that renewables may outcompete fossil fuels faster than even some experts predicted a decade ago. Biotechnological innovations—like crops engineered to withstand extreme temperatures or microorganisms designed for carbon sequestration—round out a portfolio of scientific breakthroughs that can help us keep pace with growing challenges. Still, these innovations must be balanced with caution and ethical consideration, ensuring that short-term fixes don't create new problems.

Envisioning the Future: Beyond Barriers and Towards Transformation

Chapter 8: Envisioning the Future circled back to the psychological landscape that shapes our collective response. Change, at its core, is as much about mindsets and cultural norms as it is about policy frameworks. Societies have historically resisted major shifts—whether it was the abolition of certain exploitative labor practices or the mainstreaming of women's suffrage—until a tipping point in public consciousness was reached. Climate change is no different: the seeds of transformation lie in changing hearts, minds, and daily routines on a mass scale.

This chapter also offered **Blueprints for a Green Tomorrow**, drawing on case studies where sustainability is woven into the fabric of entire communities. In these pockets of success, synergy arises between ecological health and economic development. Concepts like the circular economy show that we can decouple economic prosperity from the depletion of natural resources by reusing, repurposing, and recycling materials in closed-loop systems. **Eco-Innovators**—visionary individuals and enterprises—have shown that future industries can thrive in harmony with the planet rather than at its expense.

The concluding sections on **Cultural Shifts Towards Sustainability** underscored the growing emphasis on minimalism, conservation, and biodiversity appreciation. From local food movements that reconnect farmers and consumers to ethical fashion brands that prioritize circularity and fair labor, culture is evolving—albeit unevenly—towards an ethos that sees nature not as a commodity but as a partner in our shared existence. Meanwhile, **Sustainable Governance** is increasingly taking root in various political landscapes, hinting that the interplay between public opinion, institutional frameworks, and visionary leadership can indeed yield a more balanced relationship between human societies and Earth's ecosystems.

Climate Jobs & Opportunities

Chapter 9: Climate Jobs & Opportunities focuses on the transformative potential of climate jobs as a cornerstone of the green economy. These roles—spanning renewable energy, sustainable transportation, and eco-friendly construction—highlight how economic resilience and environmental stewardship go hand in hand. Workforce development emerges as a critical enabler, equipping individuals with the skills needed to meet the demands of this rapidly evolving landscape. The chapter emphasizes that building a **skilled workforce** is not just about addressing unemployment but about driving systemic change towards sustainability.

This chapter also explores **Fair Transitions and Inclusive Growth**, delving into how climate jobs can bridge the gap for communities historically reliant on traditional industries. By prioritizing equity through accessible training programs and targeted investments, we can mitigate the economic disruptions of the transition while fostering broad-based growth. Case studies illustrate success stories where communities have pivoted from fossil fuel dependence to renewable energy innovation. As the chapter concludes, the emphasis shifts to **Policy and Industry Collaboration**, underlining how strategic partnerships between governments, businesses, and educational institutions can pave the way for a resilient and inclusive green economy.

Climate Data—The Backbone of Understanding

Chapter 10: Climate Data is the foundation of our understanding of global change, providing the insights needed to track transformations, predict future impacts, and develop effective responses. This chapter examines the evolution of climate science, from early manual observations to cutting-edge technologies like satellites, remote sensing, and advanced climate models. It highlights how these tools reveal critical trends, such as rising sea levels and shifting weather patterns, while enabling actionable insights to mitigate the effects of climate change.

The chapter also discusses the challenges of data gaps, inconsistencies, and inequities in accessibility. Remote regions, such as the Arctic, often lack sufficient monitoring, while developing nations face significant hurdles in utilizing climate data. Collaborative initiatives like open-access satellite programs are showcased as vital steps toward bridging these divides. Ultimately, this chapter underscores that climate data is not just a tool for scientists—it is a compass for policymakers, communities, and individuals to navigate the path toward a sustainable and resilient future.

Climate Intelligence—Transforming Data into Action

Chapter 11: Climate Intelligence bridges the gap between data collection and impactful action. This chapter explores how technologies such as AI, machine learning, and predictive analytics turn raw climate data into tools for informed decision-making. From disaster preparedness to urban planning and sustainable agriculture, climate intelligence enables precise, real-time solutions that protect lives, optimize resources, and create opportunities for a more sustainable future.

The chapter also highlights real-world applications across industries, such as using AI to predict wildfires, optimize energy grids, or guide farmers on planting schedules. It delves into challenges like data accessibility, biases, and the digital divide, underscoring the importance of collaboration between governments, private enterprises, and local communities. By integrating cutting-edge technology with citizen science and indigenous knowledge, climate intelligence transforms climate action into a proactive, inclusive, and transformative force.

Moving Forward: Practical Steps and Personal Commitments

Having traversed all these chapters, the question that often arises is: "What can I do?" The closing note is incomplete without offering suggestions—concrete yet adaptable to different lifestyles, professional capacities, and socio-economic realities. Here are a few pathways:

1. **Educate and Engage**
- Continue learning about climate science, policy, and solutions. Knowledge evolves rapidly, and staying informed helps you make thoughtful decisions.

- Share what you learn with friends, family, and colleagues. Conversations—online or offline—can ripple outward and influence others.

2. **Personal Footprint and Lifestyle Choices**
- Reduce waste, conserve energy, and move toward plant-rich diets where possible.

- Opt for public transit, carpooling, or electric vehicles to lessen your transportation emissions.

- Support businesses and products that prioritize sustainability and ethical supply chains.

3. Political and Civic Action

- Vote with climate in mind and advocate for policies that foster green infrastructure, renewable energy, and equitable resource distribution.

- Engage with local government, attend council meetings, or participate in environmental committees. Policy changes at the local level can be deeply impactful.

4. Professional Influence

- If you work in a corporate or institutional setting, champion internal sustainability initiatives—anything from improved recycling programs to better supply chain practices.

- Leverage professional skills—be it in finance, law, engineering, education, or arts—to contribute solutions within your industry.

5. Community and Coalition Building

- Form or join local climate action groups, grassroots projects, or nonprofit organizations. Collective efforts amplify individual voices and can achieve more significant results.

- Reach out to multicultural or cross-sector coalitions, broadening perspectives and forging alliances that can tackle issues like climate justice more comprehensively.

6. Support for Innovation and Creativity

- Encourage and invest in research, startups, or community projects that present innovative solutions to climate adaptation and mitigation.

- Celebrate creativity—be it through art, storytelling, or design—that captures the emotional resonance of climate issues, helping to shift cultural narratives towards empathy and cooperation.

A Concluding Reflection: Embracing Our Shared Responsibility

In closing, the journey through these chapters has revealed both the depth of the climate crisis and the height of human potential. While the stakes are undeniably high, the array of solutions—from policy measures and renewable energy breakthroughs to grassroots activism and cultural shifts—is equally vast. Each chapter has woven together threads of science, sociology, politics, and innovation, underscoring that no single discipline or demographic holds all the answers. It is precisely this intersectional approach—where scientists collaborate with policymakers, urban planners consult with indigenous elders, and citizens hold corporations accountable—that will unlock the transformative potential we so urgently need.

Your role, now that you have reached the end of this book, is to carry forward this integrated perspective. The knowledge you've gained is a tool for empowerment—an invitation to question the status quo, advocate for equitable policies, support sustainable innovations, and model daily behaviors that respect planetary boundaries. Whether you are a seasoned environmental activist, a curious reader who stumbled upon these pages, or a concerned parent thinking about the future of your children, your place in this great collective effort matters.

We must remember that optimism is not a static feeling but an active choice rooted in the belief that better outcomes are possible. Armed with the insights from these chapters, you have a map of sorts—an understanding of how the climate crisis developed, the tipping points that threaten irreversible consequences, the policy frameworks and diplomatic landscapes that can either hamper or accelerate progress and the wide spectrum of solutions already taking shape. The climate challenge can often feel like an unyielding wave, but as we have explored, turning the tide is about harnessing collective momentum, each person pushing in unison to create a current of change.

The final pages of a book do not mark an ending so much as a transition—an invitation to carry its messages and insights out into the world. Climate change is arguably the defining challenge of our age, but within that challenge lies an opportunity for humanity to redefine its relationship with nature and with one another. We can envision a world that's more just, vibrant, and harmonious than anything we've known—a world where economic prosperity is no longer at odds with ecological resilience and where the shared endeavor of preserving our planet fosters unity rather than division.

Thank you for journeying through these chapters and for reflecting deeply on the issues they present. May the knowledge, strategies, and stories you've encountered energize your next steps—whatever they may be. As you close this book, take heart in the truth that each decision you make, each conversation you spark, and each community initiative you join is part of a larger, cascading movement toward a sustainable future. Our collective path forward is illuminated by creativity, empathy, perseverance, and hope. Let us walk it together, prepared to meet the challenges that lie ahead and determined to shape a world worthy of the generations that follow.

References

- Siwale, Juliana (2013). "A critical evaluation of international development and poverty: the case of microfinance and the Swedish International Development Cooperation Agency in Zambia." The University of Manchester Library. https://core.ac.uk/download/16498986.pdf

- Caprotti, F, Gong, Z (2017). "Social sustainability and residents' experiences in a new Chinese eco-city". 'Elsevier BV'. https://core.ac.uk/download/77032958.pdf

- Aktas, E, Ogunyemi, T (2013). "The impact of Green Information Systems on sustainable supply chain and organizational performance". https://core.ac.uk/download/19475151.pdf

- Blackmore, Chris, Ison, Raymond, Lane, Andrew, Reynolds, Martin (2015). "Embedding sustainability through systems thinking in practice: some experiences from the Open University". Pedagogic Research Institute and Observatory (PedRIO). https://core.ac.uk/download/82980120.pdf

- Boussauw, Kobe, Vanoutrive, Thomas (2019). "Flying green from a carbon neutral airport: the case of Brussels". 'MDPI AG'. https://core.ac.uk/download/199401222.pdf

- Dewulf, Jo, Taelman, Sue Ellen (2017). "Waste management in (peri-)urban areas: integrated assessment of environmental, social and economic sustainability in a collaborative decision support environment". https://core.ac.uk/download/287942273.pdf

- Blackmore, Chris, Ison, Raymond, Lane, Andrew, Reynolds, Martin (2015). "Embedding sustainability through systems thinking in practice: some experiences from the Open University". Pedagogic Research Institute and Observatory (PedRIO). https://core.ac.uk/download/82980120.pdf

- (2003). "ACCESS: An Inception Report". 'Association for Vascular Access'. https://core.ac.uk/download/71359459.pdf

- Charter, Martin, Clark, Tom (2007). "Sustainable innovation: key conclusions from Sustainable Innovation Conferences 2003–2006 organised by The Centre for Sustainable Design". https://core.ac.uk/download/103312.pdf

- Abdul Wahab, Aguado, Azevedo, Banawi, Bandehnezhad, Barnett-Page, Belekoukias, Besseris, Briner, Bryman, Cabral, Cabral, Cabral, Carvalho, Carvalho, Carvalho, Ceulemans, Chauhan, Chiarini, Cluzel, Darnall, Davids, Denyer, Dhingra, Dhingra, Diaz-Elsayed, Digalwar, Duarte, Duarte, Duarte, Duarte, Duarte, Dües, Esmemr, Espadinha-Cruz, Fink, Folinas, Forrester, Freyne, Galeazzo, Garza-Reyes, Garza-Reyes, Garza-Reyes, Gottberg, Govindan, Gunasekaran, Gupta, Hajmohammad, Hajmohammad, Herrmann, Herron, Hines, Hosseini, Jabbour, Johansson, Jose Arturo Garza-Reyes, Kainuma, King, Kitazawa, Kitchenham, Kleindorfer, Kurdve, Kurdve, Lambert, Larson, Lean and Green, Martinez-Jurado, Marx, Mashaei, Mason, Maxwell, Mollenkopf, Moreira, Nunes, Pampanelli, Parveen, Paumgartten, Puvanasvaran, QSR International, Ranky, Rothenberg, Rousseau, Salleh, Sarkis, Sarkis, Sarkis,

Saunders, Sawhney, Sertyesilisik, Seuring, Simpson, Slack, Smith, Sobral, Thomas, Thorpe, Tice, Tranfield, Vais, Vargo, Verrier, Wadhwa, Webster, Wiengarten, Wong, Yang, Zhu (2015). "Lean and green – a systematic review of the state of the art literature." 'Elsevier BV'. https://core.ac.uk/download/46171209.pdf

- Cuoco, Eduardo, Halberg, Niels, Huber, Machteld, Micheloni, Cristina, Niggli, Urs, Padel, Susanne, Pearce, Bruce, Schlüter, Marco, Schmid, Otto, Willer, Helga (2010). "Implementation Action Plan for organic food and farming research." Technology Platform TP organics. https://core.ac.uk/download/10931243.pdf

- Bellon, S., Desclaux, D., LE PICHON, V. (2010). "Innovation and research in organic farming: A multi-level approach to facilitate cooperation among stakeholders." Universität für Bodenkultur, Vienna. https://core.ac.uk/download/10931396.pdf

- Bansal, Neha, Shrivastava, Vineet, Singh, Jagdish (2015). "Smart Urbanization – Key to Sustainable Cities." CORP – Competence Center of Urban and Regional Planning. https://core.ac.uk/download/55284665.pdf

- Blackmore, Chris, Ison, Raymond, Lane, Andrew, Reynolds, Martin (2015). "Embedding sustainability through systems thinking in practice: some experiences from the Open University." Pedagogic Research Institute and Observatory (PedRIO). https://core.ac.uk/download/82980120.pdf

- (2003). "ACCESS: An Inception Report". 'Association for Vascular Access'. https://core.ac.uk/download/71359459.pdf

- Jesus, Eduardo Natividade, Monteiro, João Pedro Medina, Rodrigues, João Coutinho, Sousa, Nuno (2024). "Benchmarking real and ideal cities: a multicriteria analysis of city performance based on urban form." Elsevier. https://core.ac.uk/download/614510614.pdf

- Abdo, Linda, Coupland, Grey, Griffin, Sandy, Kemp, Annabeth (2019). "Biodiversity offsets can be a valuable tool in achieving sustainable development: Developing a holistic model for biodiversity offsets that incorporates environmental, social and economic aspects of sustainable development." ResearchOnline@ND. https://core.ac.uk/download/232131035.pdf

- Katz, Daniel S., Choi, Sou-Cheng T., Niemeyer, Kyle E., Hetherington, James, Löffler, Frank, Gunter, Dan, Idaszak, Ray, Brandt, Steven R., Miller, Mark A., Gesing, Sandra, Jones, Nick D., Weber, Nic, Marru, Suresh, Allen, Gabrielle, Penzenstadler, Birgit, Venters, Colin C., Davis, Ethan, Hwang, Lorraine, Todorov, Ilian, Patra, Abani, de Val-Borro, Miguel (2012). "Report on the Third Workshop on Sustainable Software for Science: Practice and Experiences (WSSSPE3)". 'Ubiquity Press, Ltd.'. https://core.ac.uk/download/74212159.pdf

- Blackmore, Chris, Ison, Raymond, Lane, Andrew, Reynolds, Martin (2015). "Embedding sustainability through systems thinking in practice: some experiences from the Open University." Pedagogic Research Institute and Observatory (PedRIO). https://core.ac.uk/download/82980120.pdf

- Jesus, Eduardo Natividade, Monteiro, João Pedro Medina, Rodrigues, João Coutinho, Sousa, Nuno (2024).

"Benchmarking real and ideal cities: a multicriteria analysis of city performance based on urban form." Elsevier. https://core.ac.uk/download/614510614.pdf

- Luchkina, Claire M. (2016). "Online Permaculture Resources: An Evaluation of a Selected Sample." ScholarWorks@UARK. https://core.ac.uk/download/84119771.pdf

- Thampy, Sarin, Valogianni, Konstantina (2023). "Do Natural and Technological approaches have any impact on Agrifarms?". AIS Electronic Library (AISeL). https://core.ac.uk/download/590878806.pdf

- Block, Thomas, Van de Velde, Riet (2016). "Transition UGent: a bottom-up initiative towards a more sustainable university." ISCN - International Sustainable Campus Network. https://core.ac.uk/download/55768294.pdf

- adureira, Ana Mafalda. "Urban Design in Neighbourhood Commodification". https://core.ac.uk/download/pdf/6633131.pdf

- Caird, Sally, Potter, Stephen, Roy, Robin (2007). "People-centred eco-design: consumer adoption of low and zero carbon products and systems." 'Informa UK Limited'. https://core.ac.uk/download/401.pdf

- Aish, Crawley, Eastman, Eastman, Eastman, Engelbart, Esther Salmerón-Manzano, Francisco Manzano-Agugliaro, Mehdi Chihib, Mishchenko, Mueller, Nawari, Ning, Nuria Novas, Rynne, Ulmanis (2019). "Bibliometric Maps of BIM and BIM in Universities: A Comparative Analysis." 'MDPI AG'. https://core.ac.uk/download/286590556.pdf

- Hemalatha J., Anjel Raj Y., Panboli S., A. L (2024). "Electric vehicle adoption toward sustainable transportation solution: key drivers and implications." https://www.semanticscholar.org/paper/78f989b48a758d6702c68f1d96dab482ba1be97c

- Katie Savin, Zachary A. Morris, Marion S. Wise, Rebecca Marinoff (2024). "Every day you are working you have to prove it": Navigating the costs of work and ableism with visual impairment." https://www.semanticscholar.org/paper/842dd4cec2f429e2ae3dfb506fb1465face47c9d

- Sayali Andhare, Namrata Dhamankar (2024). "A Wholistic Housing Solution for Onsite Construction Workers." https://www.semanticscholar.org/paper/2de9278e4b84484d833703fb6188a05920a2c48e

- Ivan Bozhikin (2023). "Trees and shrubs suitable for the construction of agroforestry systems in a temperate climate and their use in economic activities: a study in Bulgaria." https://www.semanticscholar.org/paper/4411a07b6fc162e1a34404e0aba791cae5fca7b8

- Anastasia Christman, Christine Riordan (2012). "City Systems: Building Blocks for Achieving Sustainability and Creating Good Jobs." National Employment Law Project. https://core.ac.uk/download/71361662.pdf

- Jason Walsh, Radhika Fox, Shawn Fremstad. "Bringing Home the Green Recovery: A User's Guide to the 2009 American Recovery and Reinvestment Act". https://core.ac.uk/download/pdf/6967434.pdf

- Tofikk Redi (2024). "Systematic review of mitigation approaches in Ethiopia's energy sector:

Strategies for sustainable development and climate resilience." https://www.semanticscholar.org/paper/074d595ff86db8ff5ee737400685619b977fe80b

- T. BUIH., M. PHANT., J. BABCOCKJ., Nguy (2019). "Capacity and Trade: Vietnam Oregon Initiative Shared Climate Agenda." https://www.semanticscholar.org/paper/43c56da738366b1002d6e9d6789f1c257bf181cb

- (2008). "Charting Our Own Course: Today's Challenges, Tomorrow's Opportunities, December 2008". https://core.ac.uk/download/11350328.pdf

- Sustainable Development Commission (2012). "Landscape, environment and community impacts of nuclear power." Sustainable Development Commission. https://core.ac.uk/download/1586823.pdf

- I Ilalele, Hlalele (2012). "The labor movement response to climate change: a case study of South African Transport and Allied Workers Union (SATAWU)." https://core.ac.uk/download/39670622.pdf

- Hoppe, Merja, Winter, Martin (2015). "European diversity: regional innovation potential in transportation." David Publishing. https://core.ac.uk/download/162589258.pdf

- Jacob Funk Kirkegaard, Lutz Weischer, Matt Miller, Thilo Hanemann. "Toward a Sunny Future? Global Integration in the Solar PV Industry". https://core.ac.uk/download/pdf/6552057.pdf

- Młynarski, Tomasz (2015). "Wyzwania i interesy państw wobec konieczności przeciwdziałania globalny zmianom klimatycznym". Oficyna Wydawnicza AFM. https://core.ac.uk/download/214929670.pdf

- Keen Point Consulting, TEConomy Partners (2016). "New Hampshire University Research and Industry Plan: A Roadmap for Collaboration and Innovation". University of New Hampshire Scholars\u27 Repository. https://core.ac.uk/download/215517669.pdf

- Cha, J. Mijin (2017). "A Just Transition: Why Transitioning Workers into a New Clean Energy Economy Should Be at the Center of Climate Change Policies." FLASH: The Fordham Law Archive of Scholarship and History. https://core.ac.uk/download/216957549.pdf

- Aldy, Joseph E., Pizer, William A. "The Competitiveness Impacts of Climate Change Mitigation Policies". https://core.ac.uk/download/pdf/6673675.pdf

- Joseph E. Aldy, William A. Pizer. "The Competitiveness Impacts of Climate Change Mitigation Policies". https://core.ac.uk/download/pdf/6815595.pdf

- Bruyere, Susanne M, Dr. , Filiberto, David (2013). "The Green Economy and Job Creation: Inclusion of People with Disabilities". DigitalCommons@ILR. https://core.ac.uk/download/33596583.pdf

- Cha, J. Mijin (2017). "A Just Transition: Why Transitioning Workers into a New Clean Energy Economy Should Be at the Center of Climate Change Policies." FLASH: The Fordham Law Archive of Scholarship and History. https://core.ac.uk/download/216957549.pdf

- Charles, David, Hazelkorn, E, Piacentini, M, Puukka, J, Rushford, J (2010). "OECD reviews of higher education in regional and city development, State of Victoria, Australia." OECD, Paris. https://core.ac.uk/download/9039569.pdf

- (2015). "Private Sector Investment and Sustainable Development: The Current and Potential Role of Institutional Investors, Companies, Banks and Foundations in Sustainable Development." United Nations Global Compact. https://core.ac.uk/download/75781326.pdf

- (2016). "Scaling Up Climate Action to Achieve the Sustainable Development Goals." UNDP. https://core.ac.uk/download/75760930.pdf

- Jessica Long, Nevin Vages (2013). "A Critical Scan of Four Key Topics for the Philanthropic Sector: A Study by the Rockefeller Foundation and Accenture Development Partnerships." Accenture. https://core.ac.uk/download/71362647.pdf

- (2010). "Bridging the Equity Gap: Driving Community Health Outcomes Through the Green Jobs Movement." Green for All. https://core.ac.uk/download/71345639.pdf

- Lisbeth B. Schorr, Vicky Marchand (2007). "Pathway to Successful Young Adulthood". Pathways Mapping Initiative. https://core.ac.uk/download/75786157.pdf

- Tan, Chin An (2013). "The Effect Of Different Generations Psychosocial Work Climate On The Employees' Work Performance." https://core.ac.uk/download/16515350.pdf

- Carlos Martín, Jennifer Bagnell-Stuart, Kimberly Burnett, Stephen Whitlow (2013). "Evaluation of the Sustainable

Employment in a Green US Economy (SEGUE).” Rockefeller Foundation. https://core.ac.uk/download/86445031.pdf

- (2010). “Bridging the Equity Gap: Driving Community Health Outcomes Through the Green Jobs Movement.” Green for All. https://core.ac.uk/download/71345639.pdf

- (2011). “Green Jobs in a Sustainable Food System”. Green for All. https://core.ac.uk/download/71345683.pdf

- Mark Fulton, Reid Capalino (2014). “Investing in the Clean Trillion: Closing the Clean Energy Investment Gap”. ‘Colegio de Enfermeria de Caceres’. https://core.ac.uk/download/71365113.pdf

- Christopher Flavin (2008). “Low-carbon energy: a roadmap”. Worldwatch Institute. https://core.ac.uk/download/71340916.pdf

- Boston University Institute for Sustainable Energy, Castigliego, Joshua, Cleveland, Cutler, Fortune, D’Janapha, Galante, Emma, Martin, Atyia, Perez, Taylor, Stanton, Liz, Walsh, Michael, Woods, Bryndis (2019). “Carbon Free Boston: Social equity report 2019”. Boston University Institute for Sustainable Energy. https://open.bu.edu/bitstream/2144/39229/1/CFB_Social_Equity_Report_053119.pdf

- Aquilina, Diana, Briguglio, Michael, Brown, Maria (2011). “Green jobs from a small scale perspective: case studies from Malta.” Green European Foundation. https://core.ac.uk/download/46602806.pdf

- (2015). “Private Sector Investment and Sustainable Development: The Current and Potential Role of Institutional

Investors, Companies, Banks and Foundations in Sustainable Development." United Nations Global Compact. https://core.ac.uk/download/75781326.pdf

- Carlos Martín, Jennifer Bagnell-Stuart, Kimberly Burnett, Stephen Whitlow (2013). "Evaluation of the Sustainable Employment in a Green US Economy (SEGUE)." Rockefeller Foundation. https://core.ac.uk/download/86445031.pdf

- Cha, J. Mijin (2017). "Labor Leading on Climate: A Policy Platform to Address Rising Inequality and Rising Sea Levels in New York State." DigitalCommons@Pace. https://core.ac.uk/download/84502051.pdf

- Rebecca Deehr (2016). "Oregon: Changing Climate, Economic Impacts, & Policies for Our Future". Environmental Entrepreneurs (E2). https://core.ac.uk/download/75786043.pdf

- (2009). "The Clean Energy Economy: Repowering Jobs, Businesses and Investments Across America". The Pew Charitable Trusts. https://core.ac.uk/download/71350144.pdf

- Boston University Institute for Sustainable Energy, Castigliego, Joshua, Cleveland, Cutler, Fortune, D'Janapha, Galante, Emma, Martin, Atyia, Perez, Taylor, Stanton, Liz, Walsh, Michael, Woods, Bryndis (2019). "Carbon Free Boston: Social equity report 2019". Boston University Institute for Sustainable Energy. https://open.bu.edu/bitstream/2144/39229/1/CFB_Social_Equity_Report_053119.pdf

- Henly, Megan M., Safford, Thomas G., Ulrich, Jessica D. (2011). "Jobs, natural resources, and community resilience: A survey of southeast Alaskans about social and environmental change." University of New Hampshire Scholars\u27 Repository. https://core.ac.uk/download/72047899.pdf

- Gehring, Markus, Kent, Avidan (2013). "Innovative Regulatory Frameworks Promoting Green Economy for Sustainable Development and Poverty Eradication in Europe." 'FUNEP'. https://core.ac.uk/download/41991363.pdf

- GHK Consulting, Land Use Consultants (2010). "Sustainable economic growth within environmental limits hypothetical case study: 'limitville'". East Midlands Development Agency. https://core.ac.uk/download/30624514.pdf

- Carlos Martín, Jennifer Bagnell-Stuart, Kimberly Burnett, Stephen Whitlow (2013). "Evaluation of the Sustainable Employment in a Green US Economy (SEGUE)." Rockefeller Foundation. https://core.ac.uk/download/86445031.pdf

- (2009). "The Economic Development and Workforce Development Systems". Surdna Foundation. https://core.ac.uk/download/71347509.pdf

- Carlos Martín, Jennifer Bagnell-Stuart, Kimberly Burnett, Stephen Whitlow (2013). "Evaluation of the Sustainable Employment in a Green US Economy (SEGUE)." Rockefeller Foundation. https://core.ac.uk/download/86445031.pdf

- Carlos Martín, Jennifer Bagnell-Stuart, Kimberly Burnett, Stephen Whitlow (2013). "Evaluation of the Sustainable Employment in a Green US Economy (SEGUE)." Rockefeller Foundation. https://core.ac.uk/download/86445031.pdf

- (2010). "Bridging the Equity Gap: Driving Community Health Outcomes Through the Green Jobs Movement." Green for All. https://core.ac.uk/download/71345639.pdf

- Boston University Institute for Sustainable Energy, Castigliego, Joshua, Cleveland, Cutler, Fortune, D'Janapha, Galante, Emma, Martin, Atyia, Perez, Taylor, Stanton, Liz, Walsh, Michael, Woods, Bryndis (2019). "Carbon Free Boston: Social equity report 2019". Boston University Institute for Sustainable Energy. https://open.bu.edu/bitstream/2144/39229/1/CFB_Social_Equity_Report_053119.pdf

- Atterton, Jane, Steiner, Artur (2014). "The contribution of rural businesses to community resilience." 'SAGE Publications'. https://core.ac.uk/download/293880850.pdf

- Cha, J. Mijin (2017). "Labor Leading on Climate: A Policy Platform to Address Rising Inequality and Rising Sea Levels in New York State." DigitalCommons@Pace. https://core.ac.uk/download/84502051.pdf

- Mark Fulton, Reid Capalino (2014). "Investing in the Clean Trillion: Closing the Clean Energy Investment Gap". 'Colegio de Enfermeria de Caceres'. https://core.ac.uk/download/71365113.pdf

- Buckley, Chris, Cloete, Eugene, Godfrey, Linda, Hildebrandt, Diane, Makhafola, Makhapa, Pouris, Anastassios, van Zyl, Emile, Watson, Jim (2014). "The state of green technologies in South Africa". Academy of Sciences of South Africa and Ministry of Science and Technology, Republic of South Africa. https://core.ac.uk/download/30610831.pdf

- Kryk, Barbara (2015). "Green jobs – good practices." Wydawnictwo Uniwersytetu Łódzkiego. https://core.ac.uk/download/71980475.pdf

- Gretchen Kosarko, Robert Weissbourd (2011). "Economic Impacts of GO TO 2040". Chicago Community Trust. https://core.ac.uk/download/71345969.pdf

- Andrés Cisneros-Montemayor, Gerald Singh, William Cheung, Yoshitaka Ota (2017). "Oceans and the Sustainable Development Goals: Co-Benefits, Climate Change & Social Equity." The Nippon Foundation-UBC Nereus Program. https://core.ac.uk/download/87085613.pdf

- (2008). "Charting Our Own Course: Today's Challenges, Tomorrow's Opportunities, December 2008". https://core.ac.uk/download/11350328.pdf

- Hamilton, Lisa Anne, Rábago, Karl R., Valova, Radina (2017). "Transition Support Mechanisms for Communities Facing Full or Partial Coal Power Plant Retirement in New York." DigitalCommons@Pace. https://core.ac.uk/download/84502037.pdf

- Boston University Institute for Sustainable Energy, Castigliego, Joshua, Cleveland, Cutler, Fortune, D'Janapha, Galante, Emma, Martin, Atyia, Perez, Taylor, Stanton, Liz, Walsh, Michael, Woods, Bryndis (2019). "Carbon Free Boston: Social equity report 2019". Boston University Institute for Sustainable Energy. https://open.bu.edu/bitstream/2144/39229/1/CFB_Social_Equity_Report_053119.pdf

- Gehring, Markus, Kent, Avidan (2013). "Innovative Regulatory Frameworks Promoting Green Economy for

Sustainable Development and Poverty Eradication in Europe." 'FUNEP'. https://core.ac.uk/download/41991363.pdf

- Boston University Institute for Sustainable Energy, Castigliego, Joshua, Cleveland, Cutler, Fortune, D'Janapha, Galante, Emma, Martin, Atyia, Perez, Taylor, Stanton, Liz, Walsh, Michael, Woods, Bryndis (2019). "Carbon Free Boston: Social equity report 2019". Boston University Institute for Sustainable Energy. https://open.bu.edu/bitstream/2144/39229/1/CFB_Social_Equity_Report_053119.pdf

- (2010). "Bridging the Equity Gap: Driving Community Health Outcomes Through the Green Jobs Movement." Green for All. https://core.ac.uk/download/71345639.pdf

- Pravin Tathod, Deepak Pyasi (2024). "Utilizing Digital Transformation Technology at Construction Sites to Mitigate Workplace Incidents and Enhance Safety Practices within the Construction Sector." https://www.semanticscholar.org/paper/d37a7fffce196b8b41cb825ec219876ec7dfa06

- Ifeanyi Onyedika Ekemezie, Wags Numoipiri Digitemie (2024). "CLIMATE CHANGE MITIGATION STRATEGIES IN THE OIL & GAS SECTOR: A REVIEW OF PRACTICES AND IMPACT." https://www.semanticscholar.org/paper/512a98cdc59714a3e43ce61f9926342421bb5ddb

- Lin Chen, Goodluck Msigwa, Mingyu Yang, Ahmed I. Osman, Samer Fawzy, David W. Rooney, Pow-Seng Yap (2022). "Strategies to achieve a carbon neutral society: a review." 20. pp. 2277-2310. https://doi.org/10.1007/s10311-022-01435-8

- David Rolnick, Priya L. Donti, Lynn H. Kaack, Kelly Kochanski, Alexandre Lacoste, Kris Sankaran, Andrew Slavin Ross, Nikola Milojevic-Dupont, Natasha Jaques, Anna Waldman-Brown, Alexandra Sasha Luccioni, Tegan Maharaj, Evan David Sherwin, S. Karthik Mukkavilli, Konrad P. Körding, Carla P. Gomes, Andrew Y. Ng, Demis Hassabis, John Platt, Felix Creutzig, Jennifer Chayes, Yoshua Bengio (2022). "Tackling Climate Change with Machine Learning". 55. pp. 1-96. https://doi.org/10.1145/3485128

- Dolf Gielen, Francisco Boshell, Değer Saygin, Morgan Bazilian, Nicholas Wagner, Ricardo Gorini (2019). "The role of renewable energy in the global energy transformation." 24. pp. 38-50. https://doi.org/10.1016/j.esr.2019.01.006

- (2018). "Sendai Framework for Disaster Risk Reduction 2015-2030". https://doi.org/10.1163/2210-7975_hrd-9813-2015016